BASIC MATHEMATICS

A Revision Primer for Management Students

Basic Mathematics

A Revision Primer for Management Students

Second Edition
Revised Reprint

Building Confidence

in Analytical Reasoning

and Numeracy Skills

for GMAT

TREVOR WEGNER

JUTA
AND COMPANY LTD

Basic Mathematics: A Revision Primer for Management Students
Second Edition

First published 2004
Second Edition
Reprinted 2005, 2006, 2007, 2009, 2015
Revised Second Edition 2016

Juta and Company Ltd
First Floor
Sunclare Buildong
21 Dreyer Street
Claremont
7708
PO Box 14373, Lansdowne, 7779, Cape Town, South Africa

© 2016 Juta & Company Ltd

ISBN 978-148511-383-6

Project Manager: Carlyn Bartlett-Cronje
Editor: Integrated Project Management Solutions
Proofreader: Christa Büttner-Rohwer
Cover designer: Eugene Badenhorst
Typesetter: Trace Digital Services

Designed and typeset in 10 on 12pt Stone Serif

This book is dedicated
to Shirley, Sally, Maryanne, Jessica, Melissa, Amy
and my parents.

TABLE OF CONTENTS

PREFACE

Basic Mathematics is aimed mainly at management students who intend to write the GMAT test for which a strong foundation in the fundamentals of basic mathematics is essential. It is also of value to any other student wanting a general revision of basic mathematics.

In addition to re-acquainting the student with the basic tools of mathematics, this text has two further aims: to sharpen the student's mental arithmetic skills (a skill largely lost through reliance on calculators and computers), and to introduce the student to a range of quick-and-dirty arithmetic methods to speed up one's problem-solving ability when answering GMAT-type questions.

This text provides a general revision of basic mathematical principles, rules and methods. The focus is only on those areas of mathematics required for the GMAT test. This covers four main topics: basic arithmetic, fundamental algebra, geometry and introductory statistics. This is essentially a workbook with a strong emphasis on self-practice. After a brief review of each topic's basic rules and methods, there is at least one worked example, followed by an extensive set of self-practice exercises.

This revised edition contains a few new topics in basic mathematic and fundamental algebra and is strengthened by additional GMAT-type exercises.

This text offers both a pre-revision test and a post-revision test to allow the student to benchmark his/her basic mathematical knowledge. As a refresher text, it is not intended to teach basic mathematics; instead it offers the student an opportunity to identify his/her areas of mathematical strengths and weaknesses. After working through this text and mastering the basics of mathematics, the student can expect to be more confident when attempting the maths section of the GMAT test.

Trevor Wegner
October 2015

INTRODUCTION

Purpose

The target audience for *Basic Mathematics: A Revision Primer for Management Students* is primarily prospective students of management who intend *writing the Graduate Management Assessment Test (GMAT)* as an entrance requirement for MBA studies.

It can, however, also benefit *non-GMAT students* who need to *revise basic mathematics* in order to understand mathematical terms and concepts and use basic mathematical rules to solve business-related problems.

Its purpose is threefold:
- to re-introduce a student to the *basic tools of mathematics* appropriate for GMAT;
- to build students' confidence in their *quantitative (numeracy and analytical) reasoning skills* (i.e. to re-instil confidence – which has largely been lost as a result of a dependency on a calculator or a computer – in their ability to perform mental arithmetic); and
- to introduce a student to a number of *quick-and-dirty methods* that will enable them to speed up problem solving.

By working through this text systematically and diligently, a student should feel that he/she has **consolidated his/her basic mathematical knowledge** relevant to the GMAT. He/she would also have **sharpened his/her mental arithmetic skills** (being able to perform mathematical gymnastics!) to tackle GMAT questions with confidence and from a position of mathematical strength. Since only the answer counts in GMAT, using either the rigours of mathematics or a suitable quick-and-dirty approach (whichever leads to the correct answer quickest) is acceptable.

Since no calculators are permitted in GMAT, reliance on mental arithmetic and the ability to identify and solve mathematical problems quickly and efficiently, using either accepted mathematical tools or a suitable quick-and-dirty method, are of great importance. This is where **sharpened quantitative reasoning skills** will be a strategic advantage.

A **thorough familiarity** with this material and an **increase in confidence** in his/her mathematical ability should ensure that a GMAT student does not 'stumble' over the basic underlying mathematics when answering a GMAT question. This should speed up both GMAT **problem recognition** (i.e. knowing what kinds of mathematics are being tested), e.g. 'percentages', 'ratios', 'application of Pythagoras', etc., and **problem solving** (i.e. knowing how to apply the identified method to the problem).

A GMAT candidate who has not mastered the basic mathematical tools of this text is at an immediate disadvantage when attempting the maths section of the GMAT. Since each GMAT question relies to a greater or lesser extent on some section of mathematics, it is imperative that he/she is confident in his/her ability to apply these basic mathematical tools quickly and accurately. Any hesitation or uncertainty of the mathematics reduces his/her chance of a good GMAT result.

This text is **not** intended to **teach** basic mathematics. The material is not sufficient for that purpose. Instead it is a refresher text to help each student **identify areas of mathematical strengths and weaknesses**. A student must then take the necessary steps to address these identified areas of weakness.

This text has **numerous exercises** on every section of basic mathematics. The reader is strongly encouraged to practise as many exercises as he/she feels is necessary to master each relevant section.

The **recommended approach** to adopt to assist the revision process is threefold:
1. *review and understand the rules and methods of each technique*
2. *study the worked examples to understand the application of the rules and methods*
3. *practise the exercises to reinforce this understanding.*

A set of solutions for all exercises is given in *Appendix C* as a guide for students.

Structure

The detailed content for each section follows:

Chapter 1: Basic arithmetic (understand basic arithmetic rules to develop *mental arithmetic skills*)
    Numbers and the number line (negative, zero, positive)
    Terms: integers, fractions (mixed and improper), digits, consecutive numbers, arithmetic sequences, prime numbers, absolute values, factors, multiples, quotients, divisor)
    Order of operations: BODMAS (or PEMDAS)
    Signed numbers – multiplication/division
    Relationships between numbers ($<$, $\leqslant$, $=$, $\geqslant$, $>$);
    (addition, subtraction, multiplication, division, LCM, HCF)
    Decimals (converting fractions to decimals, decimal calculations)
    Percentages (fractions as percentages; percentage increase and decrease calculations) and interest (simple and compound) calculations
    Ratios and proportions (direct and indirect)
    Indices and roots of numbers
    Speed (distance, time)
    Rates of change/flow rates

Chapter 2: Fundamental algebra (basic principles of algebra to *form and solve equations*)
    Terms: algebraic expressions, term, equation, inequality
    Formulating algebraic expressions from word problems
    Simplifying algebraic expressions (addition, subtraction, multiplication, division)
    Solving simple linear equations with one unknown
    Solving simultaneous equations (solving for two unknowns from two linear equations)
    Exponents
    Factorising (quadratics, perfect squares, difference between two squares)
    Graphs of straight line equations and functions
    Quadratic equations and the parabola graph
    Solving simultaneous inequalities using graphs
    Break-even analysis

Chapter 3: Geometry (limited to intuitive geometry required to *calculate spatial measures)*
Triangles (types and properties, angles, area)
Pythagoras' theorem
Rectangles, squares (perimeters, area)
Circles (area, circumference and arcs)
Cylinders and cuboids (box-shaped forms)

Chapter 4: Basic statistics (descriptive statistics and intuitive probabilities for *data interpretation)*
Descriptive statistics (mean, median and standard deviation)
Probability terms (outcomes, events, addition rule)
Basic probability calculations (including conditional probabilities)
Probabilities of joint events using cross-tabulation tables
Bar charts
Counting rules (factorial, combinations and permutations)
Probability trees (analyse the occurrence of sequential events)

Once the basic concepts of mathematics have been reviewed and well understood, a student must focus on *word problem solving* (i.e. learning how to translate *word problems into the appropriate mathematical way* of solving a problem). Word problems require a student to *express a real-life problem* into numeric and/or algebraic terms and then *apply suitable rules of mathematics* to solve for the unknown value(s).

Word problems are found in all four sections of mathematics, but particularly in the application areas of speed, rates, ratios, proportions, percentages (profits, discount, interest), sets (cross-tabulation tables), geometry (Pythagoras, triangles and circles) and basic statistics.

Annexures

Pre-revision test

A pre-revision test is given in *Appendix A*. Students are strongly advised to attempt this test unseen and *before* undertaking any revision to establish a benchmark of their current mathematical knowledge and analytical skills.

Post-revision test

A post-revision test is given in *Appendix B*. Students are advised to attempt this test only *after completing revision of the mathematical material* in this publication. The standard is similar to the pre-revision test and students can possibly gauge their improvement in tackling GMAT-like questions resulting from an improved understanding after working through this text.

For both the pre- and post-revision tests, a calculator may not be used. Students must calculate answers either manually or using mental arithmetic because this promotes numerical thinking and builds confidence in one's numerical skills.

CHAPTER 1

BASIC ARITHMETIC

1.1 Mental arithmetic – the beginnings

Know the following values and operations well, as they form the basis of all basic arithmetic.

1. Multiplication tables

Learn and know the following multiplication tables:

Essential learning (1 to 12 times tables)

1	2	3	4	5	6
$\boxed{1 \times 1 = 1}$	$2 \times 1 = 2$	$3 \times 1 = 3$	$4 \times 1 = 4$	$5 \times 1 = 5$	$6 \times 1 = 6$
$1 \times 2 = 2$	$\boxed{2 \times 2 = 4}$	$3 \times 2 = 6$	$4 \times 2 = 8$	$5 \times 2 = 10$	$6 \times 2 = 12$
$1 \times 3 = 3$	$2 \times 3 = 6$	$\boxed{3 \times 3 = 9}$	$4 \times 3 = 12$	$5 \times 3 = 15$	$6 \times 3 = 18$
$1 \times 4 = 4$	$2 \times 4 = 8$	$3 \times 4 = 12$	$\boxed{4 \times 4 = 16}$	$5 \times 4 = 20$	$6 \times 4 = 24$
$1 \times 5 = 5$	$2 \times 5 = 10$	$3 \times 5 = 15$	$4 \times 5 = 20$	$\boxed{5 \times 5 = 25}$	$6 \times 5 = 30$
$1 \times 6 = 6$	$2 \times 6 = 12$	$3 \times 6 = 18$	$4 \times 6 = 24$	$5 \times 6 = 30$	$\boxed{6 \times 6 = 36}$
$1 \times 7 = 7$	$2 \times 7 = 14$	$3 \times 7 = 21$	$4 \times 7 = 28$	$5 \times 7 = 35$	$6 \times 7 = 42$
$1 \times 8 = 8$	$2 \times 8 = 16$	$3 \times 8 = 24$	$4 \times 8 = 32$	$5 \times 8 = 40$	$6 \times 8 = 48$
$1 \times 9 = 9$	$2 \times 9 = 18$	$3 \times 9 = 27$	$4 \times 9 = 36$	$5 \times 9 = 45$	$6 \times 9 = 54$
$1 \times 10 = 10$	$2 \times 10 = 20$	$3 \times 10 = 30$	$4 \times 10 = 40$	$5 \times 10 = 50$	$6 \times 10 = 60$
$1 \times 11 = 11$	$2 \times 11 = 22$	$3 \times 11 = 33$	$4 \times 11 = 44$	$5 \times 11 = 55$	$6 \times 11 = 66$
$1 \times 12 = 12$	$2 \times 12 = 24$	$3 \times 12 = 36$	$4 \times 12 = 48$	$5 \times 12 = 60$	$6 \times 12 = 72$

7	8	9	10	11	12
$7 \times 1 = 7$	$8 \times 1 = 8$	$9 \times 1 = 9$	$10 \times 1 = 10$	$11 \times 1 = 11$	$12 \times 1 = 12$
$7 \times 2 = 14$	$8 \times 2 = 16$	$9 \times 2 = 18$	$10 \times 2 = 20$	$11 \times 2 = 22$	$12 \times 2 = 24$
$7 \times 3 = 21$	$8 \times 3 = 24$	$9 \times 3 = 27$	$10 \times 3 = 30$	$11 \times 3 = 33$	$12 \times 3 = 36$
$7 \times 4 = 28$	$8 \times 4 = 32$	$9 \times 4 = 36$	$10 \times 4 = 40$	$11 \times 4 = 44$	$12 \times 4 = 48$
$7 \times 5 = 35$	$8 \times 5 = 40$	$9 \times 5 = 45$	$10 \times 5 = 50$	$11 \times 5 = 55$	$12 \times 5 = 60$
$7 \times 6 = 42$	$8 \times 6 = 48$	$9 \times 6 = 54$	$10 \times 6 = 60$	$11 \times 6 = 66$	$12 \times 6 = 72$
$\boxed{7 \times 7 = 49}$	$8 \times 7 = 56$	$9 \times 7 = 63$	$10 \times 7 = 70$	$11 \times 7 = 77$	$12 \times 7 = 84$
$7 \times 8 = 56$	$\boxed{8 \times 8 = 64}$	$9 \times 8 = 72$	$10 \times 8 = 80$	$11 \times 8 = 88$	$12 \times 8 = 96$
$7 \times 9 = 63$	$8 \times 9 = 72$	$\boxed{9 \times 9 = 81}$	$10 \times 9 = 90$	$11 \times 9 = 99$	$12 \times 9 = 108$
$7 \times 10 = 70$	$8 \times 10 = 80$	$9 \times 10 = 90$	$\boxed{10 \times 10 = 100}$	$11 \times 10 = 110$	$12 \times 10 = 120$
$7 \times 11 = 77$	$8 \times 11 = 88$	$9 \times 11 = 99$	$10 \times 11 = 110$	$\boxed{11 \times 11 = 121}$	$12 \times 11 = 132$
$7 \times 12 = 84$	$8 \times 12 = 96$	$9 \times 12 = 108$	$10 \times 12 = 120$	$11 \times 12 = 132$	$\boxed{12 \times 12 = 144}$

Note: The bold, boxed numbers are perfect squares.

Desirable learning (13 to 20 times tables)

13	14	15	16
$13 \times 1 = 13$	$14 \times 1 = 14$	$15 \times 1 = 15$	$16 \times 1 = 16$
$13 \times 2 = 26$	$14 \times 2 = 28$	$15 \times 2 = 30$	$16 \times 2 = 32$
$13 \times 3 = 39$	$14 \times 3 = 42$	$15 \times 3 = 45$	$16 \times 3 = 48$
$13 \times 4 = 52$	$14 \times 4 = 56$	$15 \times 4 = 60$	$16 \times 4 = 64$
$13 \times 5 = 65$	$14 \times 5 = 70$	$15 \times 5 = 75$	$16 \times 5 = 80$
$13 \times 6 = 78$	$14 \times 6 = 84$	$15 \times 6 = 90$	$16 \times 6 = 96$
$13 \times 7 = 91$	$14 \times 7 = 98$	$15 \times 7 = 105$	$16 \times 7 = 112$
$13 \times 8 = 104$	$14 \times 8 = 112$	$15 \times 8 = 120$	$16 \times 8 = 128$
$13 \times 9 = 117$	$14 \times 9 = 126$	$15 \times 9 = 135$	$16 \times 9 = 144$
$13 \times 10 = 130$	$14 \times 10 = 140$	$15 \times 10 = 150$	$16 \times 10 = 160$
$13 \times 11 = 143$	$14 \times 11 = 154$	$15 \times 11 = 165$	$16 \times 11 = 176$
$13 \times 12 = 156$	$14 \times 12 = 168$	$15 \times 12 = 180$	$16 \times 12 = 192$
$\boxed{13 \times 13 = 169}$	$14 \times 13 = 182$	$15 \times 13 = 195$	$16 \times 13 = 208$
	$\boxed{14 \times 14 = 196}$	$15 \times 14 = 210$	$16 \times 14 = 224$
		$\boxed{15 \times 15 = 225}$	$16 \times 15 = 240$
			$\boxed{16 \times 16 = 256}$

17	18	19	20
$17 \times 1 = 17$	$18 \times 1 = 18$	$19 \times 1 = 19$	$20 \times 1 = 20$
$17 \times 2 = 34$	$18 \times 2 = 36$	$19 \times 2 = 38$	$20 \times 2 = 40$
$17 \times 3 = 51$	$18 \times 3 = 54$	$19 \times 3 = 57$	$20 \times 3 = 60$
$17 \times 4 = 68$	$18 \times 4 = 72$	$19 \times 4 = 76$	$20 \times 4 = 80$
$17 \times 5 = 85$	$18 \times 5 = 90$	$19 \times 5 = 95$	$20 \times 5 = 100$
$17 \times 6 = 102$	$18 \times 6 = 108$	$19 \times 6 = 114$	$20 \times 6 = 120$
$17 \times 7 = 119$	$18 \times 7 = 126$	$19 \times 7 = 133$	$20 \times 7 = 140$
$17 \times 8 = 136$	$18 \times 8 = 144$	$19 \times 8 = 152$	$20 \times 8 = 160$
$17 \times 9 = 153$	$18 \times 9 = 162$	$19 \times 9 = 171$	$20 \times 9 = 180$
$17 \times 10 = 170$	$18 \times 10 = 180$	$19 \times 10 = 190$	$20 \times 10 = 200$
$17 \times 11 = 187$	$18 \times 11 = 198$	$19 \times 11 = 209$	$20 \times 11 = 220$
$17 \times 12 = 204$	$18 \times 12 = 216$	$19 \times 12 = 228$	$20 \times 12 = 240$
$17 \times 13 = 221$	$18 \times 13 = 234$	$19 \times 13 = 247$	$20 \times 13 = 260$
$17 \times 14 = 238$	$18 \times 14 = 252$	$19 \times 14 = 266$	$20 \times 14 = 280$
$17 \times 15 = 255$	$18 \times 15 = 270$	$19 \times 15 = 285$	$20 \times 15 = 300$
$17 \times 16 = 272$	$18 \times 16 = 288$	$19 \times 16 = 304$	$20 \times 16 = 320$
$\boxed{17 \times 17 = 289}$	$18 \times 17 = 306$	$19 \times 17 = 323$	$20 \times 17 = 340$
	$\boxed{18 \times 18 = 324}$	$19 \times 18 = 342$	$20 \times 18 = 360$
		$\boxed{19 \times 19 = 361}$	$20 \times 19 = 380$
			$\boxed{20 \times 20 = 400}$

2. Squares

Learn and know the squares of all numbers from 1 to 20, 25, 30, 40, 50, 60, 70, 80, 90 and 100:

$1^2 = 1$	$11^2 = 121$	$25^2 = 625$
$2^2 = 4$	$12^2 = 144$	$30^2 = 900$
$3^2 = 9$	$13^2 = 169$	$40^2 = 1\ 600$
$4^2 = 16$	$14^2 = 196$	$50^2 = 2\ 500$
$5^2 = 25$	$15^2 = 225$	$60^2 = 3\ 600$
$6^2 = 36$	$16^2 = 256$	$70^2 = 4\ 900$
$7^2 = 49$	$17^2 = 289$	$80^2 = 6\ 400$
$8^2 = 64$	$18^2 = 324$	$90^2 = 8\ 100$
$9^2 = 81$	$19^2 = 361$	$100^2 = 10\ 000$
$10^2 = 100$	$20^2 = 400$	

3. Square roots

Learn and know the square root of all square integers up to 20, e.g. $\sqrt{256} = 16$; $\sqrt{324} = 18$. In other words, know the square numbers on page 2 in reverse in order to answer questions such as:

'what is the square root of 81?'   (answer = 9)
'what is the square root of 256?'   (answer = 16)
'what is the square root of 3 600?'   (answer = 60)

To **simplify a square root**, split it into two factors, of which one is a perfect square.

Example 1	**Example 2**
Simplify $\sqrt{50}$.	Simplify $\sqrt{96}$.
Worked solution	**Worked solution**
Re-write $\sqrt{50}$ as:	Re-write $\sqrt{96}$ as:
$= \sqrt{(25 \times 2)} = \sqrt{25} \times \sqrt{2} = 5\sqrt{2}$	$= \sqrt{(16 \times 6)} = \sqrt{16} \times \sqrt{6} = 4\sqrt{6}$

4. Powers of base 2

Know all powers of 2 from 2^0 to 2^{10}.

$2^0 = 1$	$2^1 = 2$	$2^2 = 4$	$2^3 = 8$	$2^4 = 16$
$2^5 = 32$	$2^6 = 64$	$2^7 = 128$	$2^8 = 256$	$2^9 = 512$

$2^{10} = 1\ 024$ (known as a kilobyte in computers)

Example 1

A population of bacteria doubles every 15 minutes. If the population consists of 20 bacteria in the beginning, how large is the population of bacteria after one hour?

Worked solution

The population will double 4 times in one hour $\left(\frac{60}{15} = 4\right)$, thus the population grows by a factor of 2^4 (i.e. $2 \times 2 \times 2 \times 2$) in one hour.
The size after one hour is:
$20 \times 2^4 = 20 \times 16 = 320$ bacteria

Example 2

A piece of string 48 cm long is repeatedly cut in half n times.
If $n = 3$, what is the length of each piece of cut string?

Worked solution

Divide 48 by 2 three times $\frac{48}{(2 \times 2 \times 2)} = \frac{48}{2^3} = 6$ cm

The length of each cut string is 6 cm.

5. Powers of base 3

Learn and know all powers of 3 from 3^0 to 3^4.

$3^0 = 1$ $3^1 = 3$ $3^2 = 9$ $3^3 = 27$ $3^4 = 81$

6. Powers of base 10

Learn and know all powers of 10 from 10^0 to 10^6.

$10^0 = 1$ $10^1 = 10$ $10^2 = 100$ $10^3 = 1\ 000$

$10^4 = 10\ 000$ $10^5 = 100\ 000$ $10^6 = 1\ 000\ 000$ (million)

Note: For the powers of 10, the power identifies the number of zeros after the 1.

Example 1	Example 2
Write 8 000 in powers of bases 2 and 10.	Express 0.0256 in powers of bases 2 and 10.
Worked solution	**Worked solution**
$8\ 000 = 8 \times 1\ 000$ $= 2^3 \times 10^3$	$\frac{256}{10\ 000} = \frac{2^8}{10^4}$

Example 3

Write 81 000 in powers of bases 3 and 10.

Worked solution

$81 \times 1\ 000 = 3^4 \times 10^3$

7. Fractions as decimals and percentages

Learn and know certain fractions as decimal fractions and percentages.

$\frac{1}{3} = 0.333$ (33.33%) $\frac{2}{3} = 0.667$ (66.67%)

$\frac{1}{4} = 0.25$ (25%) $\frac{1}{2} = 0.5$ (50%) $\frac{3}{4} = 0.75$ (75%)

$\frac{1}{5} = 0.2$ (20%) $\frac{2}{5} = 0.4$ (40%) $\frac{3}{5} = 0.6$ (60%) $\frac{4}{5} = 0.8$ (80%)

$\frac{1}{8} = 0.125$ (12.5%) $\frac{3}{8} = 0.375$ (37.5%) $\frac{5}{8} = 0.625$ (62.5%) $\frac{7}{8} = 0.875$ (87.5%)

Exercise 1

1. Simplify:
   (a) $\sqrt{48}$ (b) $\sqrt{54}$ (c) $\sqrt{108}$ (d) $\sqrt{63}$ (e) $\sqrt{88}$
   (f) $\sqrt{396}$ (g) $\sqrt{675}$ (h) $\sqrt{588}$ (i) $\sqrt{726}$ (j) $\sqrt{60}$ (k) $\sqrt{45}$

2. Express each number in powers of bases of 2, 3 and 10 only:
   (a) 1 600   (b) 900   (c) 12 000   (d) 180 000
   (e) 1.28   (f) 0.081   (g) 0.0009   (h) 25.6

3. (a) A population of a bacteria doubles every 3 minutes. How many minutes will it take to grow from 100 to 25 600?

   (b) A population of bacteria doubles every 10 minutes. If the population size was initially 16, what would its size be in one hour?
   (Express the answer in index form.)

   (c) Four hours from now, the population of a bacteria colony will be 1.28×10^6. If the population doubles every 4 hours, what was the population size 12 hours ago? (Keep the answer in index form.)

1.2 Numbers

Basic arithmetic manipulates *numbers*. It is therefore important to understand the number line, the relationship of numbers to each other, the different types of numbers and the basic laws of numbers.

Numbers are made up of the *digits* 0, 1, 2, 3, 4, 5, 6, 7, 8 and 9. A digit is identified by its position in the number relative to the decimal point.

Example

The number 738.24 consists of 5 digits: 8 is the *units* digit; 3 is the *tens* digit; and 7 is the *hundreds* digit.
In the decimal positions, 2 is the *tenths* digit; and 4 is the *hundredths* digit.

When a number is *multiplied* by 10, all digits move *up* one position from the decimal point in the new number (i.e. the units digit becomes a tens digit; the tens digit becomes the hundreds digit, etc.). Similarly, when a number is *divided* by 10, all digits move *down* by one position.

Example

$384.75 \times 10 = 3\ 847.5$. The *tenths* digit of 384.75 (i.e. 7) becomes the *units* digit of 3 847.5; the *units* digit of 384.75 (i.e. 4) becomes the *tens* digit of 3 847.5; the *tens* digit of 384.75 (i.e. 8) becomes the *hundreds* digit of 3 847.5; etc.

$246.89 \div 10 = 24.689$. The *tens* digit of 246.89 (i.e. 4) becomes the *units* digit of 24.689; the *units* digit of 246.89 (i.e. 6) becomes the *tenths* digit of 24.689; etc.

1. The number line

The *number line* is made up of all *real numbers* from minus infinity $(-\infty)$ to plus infinity $(+\infty)$. It consists of *negative numbers* (i.e. less than zero), *zero* and *positive numbers* (i.e. greater than zero).

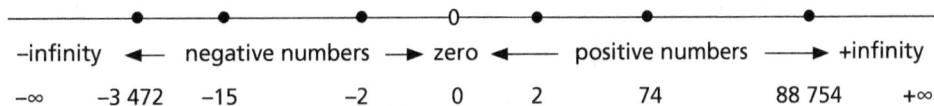

–infinity ← negative numbers → zero ← positive numbers → +infinity

| $-\infty$ | $-3\ 472$ | -15 | -2 | 0 | 2 | 74 | $88\ 754$ | $+\infty$ |

Order on the number line

Numbers to the left are always smaller than numbers to the right. Therefore, if we know that $a > b$, where a and b are any two real numbers, than a is to the right of b on the number line. Similarly, if $a < b$, then a is to the left of b on the number line.

Exercise 2

1. Which of the following relationships are true and which are false?

   (a) $7 > 12$ (b) $-4 > -6$ (c) $-3 > 0$ (d) $13 > -5$

   (e) $0 < -8$ (f) $6 < 10$ (g) $1\frac{1}{4} > 1.25$ (h) $-7 > 5$

   (i) $-4 = 4$ (j) $6 > 4$ (k) $9 < 11$ (l) $-7.4 > -7\frac{2}{5}$

   (m) $0 > -5$ (n) $-8 > 2$ (o) $-6 < -4$ (p) $-8 > 0$

   (q) $-2\frac{1}{2} < -3\frac{1}{2}$ (r) $0 < -9$

2. For each relationship below, draw a number line graph and shade in the set of numbers (called the **domain**) that satisfies it.

   (a) $x > 2$ (b) $x < -4$ (c) $x \geq -3.5$ (d) $-3 < x < 0$

   (e) $x \leq 5$ (f) $-4 < x \leq 0$ (g) $6\frac{1}{2} > x > -3$ (h) $0 \leq x \leq 6$

   (i) $-7 \geq x \geq -11$ (j) $x = 2$ (k) $x \geq -4$ (l) $x < 0$

   (m) $-7 < x \leq -5$ (n) $x \geq 1\frac{1}{2}$ (o) $x \neq -2$ (p) $x \geq -1$

   (q) $x \neq 0$ (r) $-3 \leq x < 0$ (s) $-1 \leq x < 3\frac{1}{4}$ (t) $4 > x \geq -1$

   (u) $-6.5 < x \leq -4.5$

   Hint: Substitute real numbers, if necessary, to test for the correct answers.

3. If $k > l$, $m > n$, $p > l$ and $l > m$, which of the following must be true?

   (a) $k > p$ (b) $p > n$ (c) $k > m$

4. If it is true that $x > -2$ and $x < 7$, which of the following are valid statements?

   (a) $x > 2$ (b) $x > -7$ (c) $x < 2$ (d) $-7 < x < 2$

5. If $x < 5$ and $y < 3$, is xy always less than 15?

6. For the numbers shown on the number line, which one of the following statements is true?

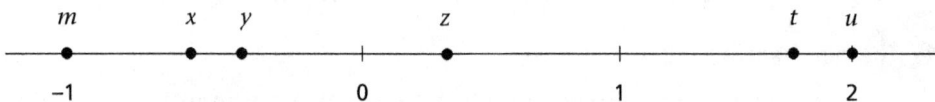

   (a) $x + y > y$ (b) $xz > xy$ (c) $tz < z$ (d) $m^u < 0$ (e) $t^m > 0$

7. On the number line, if $r < s$, p is halfway between r and s, and t is one-third of the distance between r and p from r, what is the value of $\frac{(s-t)}{(p-t)}$?

8. If $p < q$, $s > t$, $q > s$ and $t > p$, put the four numbers in ascending order.

9. How many numbers between 3 200 and 4 600 have a tens digit of 1 and a units digit of 8?

10. How many two-digit numbers are there that do not contain the digits 3 or 5?

11. Which of the following statements identify the tens digit of a number, r?

   (a) the tens digit of $\frac{r}{10}$ is 3   (b) the hundreds digit of $10r$ is 6

   (c) the tens digit of $r + 8$ is 5

12. If x is a positive number, and if the units digit of x^2 is 9, and the units digit of $(x + 1)^2$ is 4, what is the units digit of $(x + 2)^2$?

13. What is the 29th digit after the decimal point of the decimal equivalent of $\frac{7}{11}$?

2. Types of numbers

Numbers are either *integers* (whole numbers) or *fractions* (part of a whole number).

a. Integers (whole numbers)

An **integer** is any whole number (e.g. 176, –3, 0, –85, 1 633).

An integer can be *positive* (e.g. 1, 8, 154, 1 865) or *negative* (e.g. –464, –94, –1).

Note: *Zero* (0) is also an integer, but it is neither positive nor negative (it is neutral).

An integer can be *even* (–22, –4, 66, 102, etc.) or *odd* (–9, –1, 5, 7, 19, etc.)

Zero (0) is an *even* integer (as it belongs to the sequence of even integers (e.g. –6, –4, –2, **0**, 2, 4, 6, …).

When odd (O) and even (E) integers are added or subtracted, the following results will always occur:

O + O = E (e.g. 3 + 7 = 10)   O + E = O (e.g. 5 + 8 = 13) and E + E = E (e.g. 4 + 8 = 12)

When odd (O) and even (E) integers are multiplied, the following results will always occur:

O × O = O (e.g. 3 × 7 = 21)   O × E = E (e.g. 5 × 8 = 40) and E × E = E (e.g. 6 × 8 = 48)

Consecutive integers are integers that follow each other sequentially by adding 1 to the previous integer.

Examples
Five consecutive integers starting at 14 are: 14, 15, 16, 17 and 18.
Five consecutive integers starting at –3 are: –3, –2, –1, 0 and 1.

In general, the following formula applies to consecutive integers, where n is any integer (positive, zero or negative): $n, (n + 1), (n + 2), (n + 3), (n + 4)$, etc.

- Consecutive *positive* integers include only positive integers (e.g. from 1, 2, 3, 4, etc.).
- Consecutive *negative* integers include only negative integers (e.g. –4, –3, –2, to –1).
- Consecutive *odd* integers include only sequential odd integers (e.g. 7, 9, 11, 13, 15, etc. or –9, –7, –5, –3, –1, 1, 3, 5, etc.).

- Consecutive *even* integers include only sequential even integers (e.g. 24, 26, 28, 30, etc. or –10, –8, –6, –4, –2, 0, 2, 4, 6, etc.).

Arithmetic sequences

A sequence is a set of integers separated by a common difference (e.g. 5, 9, 13, 17, …). 5 is the *first term* of this sequence with a *common difference* of 4.

The formula $a_n = a_1 + (n – 1)d$ can be used to find any term of an arithmetic sequence.
a_1 is the *first term* of the sequence and d is the *common difference*.

Example

Find the 9th term of the sequence 4, 7, 10, 13, 16, …
$a_1 = 4$, $d = 3$ and $n = 9$, so $a_9 = 4 + (9 – 1)3 = 28$. The 9th term is 28.

The formula $S_n = \dfrac{n(a_1 + a_n)}{2}$ can be used to find the sum of terms in a sequence.

Example

Find the sum of the first 7 terms of the sequence 9, 12, 15, 18, 21, 24 and 27.
$a_1 = 9$, $a_7 = 27$, $n = 7$, so $S_7 = \dfrac{7(9 + 27)}{2} = 126$.

Exercise 3

1. List seven consecutive *positive* integers, starting with:

   (a) 5 (b) 28 (c) 81 (d) 302

2. List seven consecutive *negative* integers, starting with:

   (a) –12 (b) –33 (c) –40 (d) –102

3. List seven consecutive *odd* integers, starting with:

   (a) 17 (b) 3 (c) –7 (d) –21

4. List seven consecutive *even* integers, starting with:

   (a) –2 (b) 18 (c) –60 (d) –6

5. If x, y and z are consecutive positive integers and $x < y < z$, which of the following must be true?

   (a) $z – x = 2$ (b) $x.y.z$ is an even integer (c) $\dfrac{x + y + z}{3}$ is an integer

6. If x and y are integers and $–2 < x \le 3$ and $y > 2$, which of the following can never be true for xy?

   (a) –6 (b) –4 (c) –1 (d) 0 (e) 4

7. The average of five *consecutive* integers is an odd number. Which of the following statements *must* be true? (More than one can be true.)

   (a) The largest of the integers is even

   (b) The sum of the integers is odd

   (c) The difference between the largest integer and the smallest integer is an even number

8. (a) Find the 7th term of the sequence 11, 14, 17, 20, ...

    (b) Find the 8th term of the sequence –9, –4, 1, 6, ...

    (c) Find the 11th term of the sequence 8, 9.5, 11, 12.5, 14, ...

9. Find the sum of the terms in the following sequences:

    (a) 8, 11, 14, 17, 20, 23, 26, 29, 32

    (b) –10, –6, –2, 2, 6, 10, 14, 18

    (c) 3.5, 5.5, 7.5, 9.5, 11.5, 13.5

10. If positive integers x and y are not both odd, which one of the following must be even?

    (a) xy (b) $x + y$ (c) $x - y$ (d) $x + y - 1$ (e) $2(x + y) - 1$

b. Fractions

A *fraction* is a part of a whole number (e.g. $\frac{3}{5}; -\frac{4}{7}; \frac{9}{16}; \frac{3}{4}; -\frac{8}{11}$)

The *top number* is called the *numerator*. The *bottom number* is called the *denominator*.

For example, in the fraction $\frac{3}{5}$, the numerator is 3 and the denominator is 5.

Number line of integers and fractions

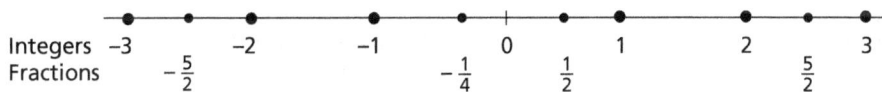

Integers	–3		–2		–1	0	1	2		3
Fractions		$-\frac{5}{2}$			$-\frac{1}{4}$	$\frac{1}{2}$		$\frac{5}{2}$		

The rules of arithmetic with fractions (i.e. add, subtract, multiply and divide) are covered in part 1.4.

c. Zero: laws of zero

For any number x ($x \neq 0$), the following rules apply:

Addition and subtraction

$x + 0 = x$ $x - 0 = x$ The answer remains unchanged.

e.g. $12 + 0 = 12$ $7 - 0 = 7$ $-16 + 0 = -16$

Multiplication

$x \times 0 = 0$ Any number multiplied by zero is always zero.

e.g. $8 \times 0 = 0$ $946 \times 0 = 0$ $-28 \times 0 = 0$

Division

$\frac{0}{x} = 0$ Any number *divided into* zero is always *zero*.

e.g. $\frac{0}{6} = 0$ $\frac{0}{(-12)} = 0$ $\frac{0}{88} = 0$

Note: $\frac{x}{0}$ is undefined. Any number *divided by* zero is *undefined* (it has no meaning).

e.g. $\frac{18}{0}$ is undefined $-\frac{562}{0}$ is undefined $\frac{0}{0}$ is undefined. It is not 1!

Three properties of zero:

    (i) Zero is an *even integer*.

    (ii) Zero is neither positive nor negative (it is a neutral number).

    (iii) Zero is not a prime number (see prime numbers below.)

When to include zero or exclude zero in the set of integers, x

- *Exclude zero* if x is defined as the set of all *positive* integers {1, 2, 3, 4, 5, ...} or as the set of all *negative* integers {..., −5, −4, −3, −2, −1}.

- *Include zero* if x is defined as the set of all *non-negative* integers {0, 1, 2, 3, 4 ...} or as the set of all *non-positive* integers {..., −5, −4, −3, −2, −1, 0}.

3. Absolute values (written as |number|)

The absolute value of any number, x, is written as $|x|$.

An *absolute value* is a measure of the actual physical *distance* between a *number* and *zero* on the number line, regardless of whether the number is negative or positive. It is therefore always a *positive value*.

e.g. The number 4 and −4 have the same absolute value, written as $|4|$, as they both lie 4 units from zero. (The + sign is never written in the answer).

Example 1	**Example 2**
What is the absolute value of $|+7|$?	What is the absolute value of $|-19-4|$?
Worked solution	**Worked solution**
$|+7| = 7$	$|-19-4| = |-23| = 23$

Example 3	**Example 4**
What is the absolute value of $|2\frac{2}{3}|$?	What is the absolute value of $|-14| - |6|$?
Worked solution	**Worked solution**
$|2\frac{2}{3}| = 2\frac{2}{3}$	$|-14| - |6| = 14 - 6 = 8$

4. Reciprocals

The **reciprocal** of a number is its *inverse*,
e.g. the reciprocal of 4 is $\frac{1}{4}$; the reciprocal of $\frac{5}{3}$ is $\frac{3}{5}$.

A **rule** to test for reciprocals is:
If the *product of two numbers* is 1, then these numbers are reciprocals.

Example	**Worked solution**
Are $\frac{\sqrt{3}}{3}$ and $\sqrt{3}$ reciprocals?	$\frac{\sqrt{3}}{3} \times \frac{\sqrt{3}}{1}$
	$= \frac{(\sqrt{3} \times \sqrt{3})}{(3 \times 1)} = \frac{(\sqrt{3})^2}{3} = \frac{3}{3} = 1$
	Yes, they are reciprocals.

Exercise 4

1. Solve the following:

   (a) $|+24|$ (b) $|+6.25|$ (c) $|-\frac{5}{8}|$ (d) $|(+9-4)|$

   (e) $|(+3-16)|$ (f) $|\left(\frac{1}{2}-2\frac{1}{2}\right)|$ (g) $|2|+|8|$ (h) $|-3|+|-10|$

   (i) $|-14|-|10|$ (j) $|5|-|-12|$ (k) $-|-18|-|-12|$ (l) $-|2|+|-7|$

2. Which of these coordinates has the greatest absolute value?

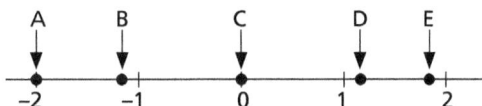

3. If $-5 < x < 0$ and $0 < y < 6$, which *one* of the following is true for $|xy|$?

   (a) $|xy| < -30$ (b) $-30 < |xy| < 30$ (c) $0 < |xy| < 30$ (d) $|xy| = 30$

4. Which of the following pairs of numbers are reciprocals?

   (a) $\frac{4}{9}$ and $\frac{27}{12}$ (b) $\frac{2}{7}$ and $-\frac{7}{2}$ (c) $3\sqrt{2}$ and $\frac{\sqrt{2}}{6}$ (d) $\frac{8}{\sqrt{3}}$ and $\frac{3}{\sqrt{8}}$

1.3   Arithmetic operations on numbers

1. Adding and subtracting numbers

When *adding positive* numbers, the answer will move in the *positive direction* on the number line (e.g. $+8 + 6 = +14$)

When *adding negative* numbers, the answer will move in the *negative direction* on the number line (e.g. $-5 - 7 = -12$)

When *subtracting positive* numbers, the answer will move in the *negative direction* on the number line (e.g. $+12 - (+9) = 3$)

When *subtracting negative* numbers the answer will move in the *positive direction* on the number line (e.g. $-6 - (-4) = -6 + 4 = -2$).

Example 1	**Example 2**
Calculate $+12 - 4$.	Calculate $-8 - 5$.
Worked solution	**Worked solution**
$+12 - 4 = +8$	$-8 - 5 = -13$

Example 3	**Example 4**
Calculate $+7 - 11$.	Calculate $-6 + 15$.
Worked solution	**Worked solution**
$+7 - 11 = -4$	$-6 + 15 = +9$

Example 5	Example 6
Calculate 0 – 3.	Calculate +9 – 12.
Worked solution	**Worked solution**
$0 - 3 = -3$	$+9 - 12 = -3$

2. Multiplying and dividing with negative numbers

Rule 1   When the two numbers have *like signs* (either both positive, or both negative), the answer is always a *positive* number.

Multiplication

$$(+a) \times (+b) = +(ab)$$

e.g. $(+13) \times (+8) = +104$

$$(-a) \times (-b) = +(ab)$$

e.g. $(-7) \times (-5) = +35$

Division

$$\frac{(+a)}{(+b)} = +\left(\frac{a}{b}\right) \qquad \text{both positive}$$

$$\frac{(+24)}{(+6)} = +4$$

$$\frac{(-a)}{(-b)} = +\left(\frac{a}{b}\right) \qquad \text{both negative}$$

$$\frac{(-72)}{(-8)} = +9$$

Rule 2   When the two numbers have *unlike signs* (one negative and one positive), the answer is always a *negative* number.

Multiplication

$$(+a) \times (-b) = -(ab)$$

e.g. $(+21) \times (-5) = -105$

$$(-a) \times (+b) = -(ab)$$

e.g. $(-13) \times (+3) = -39$

Division

$$\frac{(+a)}{(-b)} = -\left(\frac{a}{b}\right) \qquad \text{one positive, one negative}$$

$$\frac{(-72)}{(+8)} = -9$$

$$\frac{(-a)}{(+b)} = -\left(\frac{a}{b}\right) \qquad \text{one negative, one positive}$$

$$\frac{(-20)}{(+4)} = -5$$

To summarise: Like signs give positive answers while unlike signs give negative answers.

Example 1	Example 2
Calculate -3×-4.	Calculate $+5 \times (+7)$.
Worked solution	**Worked solution**
$-3 \times -4 = 12$	$+5 \times (+7) = +35$

Example 3	Example 4
Calculate $\frac{-18}{(-6)}$.	Calculate $+5 \times (-6)$.
Worked solution	**Worked solution**
$\frac{-18}{(-6)} = +3$	$+5 \times (-6) = -30$

Example 5

Calculate $\frac{-21}{(+3)}$.

Worked solution

$\frac{-21}{(+3)} = -7$

Example 6

Calculate $\frac{24}{(-8)}$.

Worked solution

$\frac{24}{(-8)} = -3$

3. Order of arithmetic operations

Always perform arithmetic operations in the following order: (BODMAS)
– **B**rackets (**o**f – meaning multiply)   BO
– **D**ivision and **M**ultiplication; and finally   DM
– **A**ddition and **S**ubtraction.   AS

Basic rule: Always work out × and ÷ before + and –.

Another acronym commonly used is PEMDAS. It stands for the same order of operations.

P (Parentheses)   E (Exponents)   M (Multiplication)
D (Division)   A (Addition)   S (Subtraction)

Example 1

Calculate $7 \times 4 + 2(9 - 6)$.

Worked solution

$7 \times 4 + 2(9 - 6)$
$= 7 \times 4 + 2(3)$
$= 28 + 6$
$= 34$

Example 2

Calculate $12 - 8(2 \times 5) + 2\left(\frac{15}{5}\right)$.

Worked solution

$12 - 8(2 \times 5) + 2\left(\frac{15}{5}\right)$.
$= 12 - 8(10) + 2(3)$
$= -62$

Example 3

Calculate $10 + \frac{12}{3} - 7 \times 2$.

Worked solution

$10 + \frac{12}{3} - 7 \times 2 = 10 + 4 - 14 = 0$

Exercise 5

1. Simplify. (Do these calculations mentally, if possible!)

(a)	$10 + 14$	(b)	$-18 + 7$	(c)	$-27 + 9$	(d)	$-11 + 11$
(e)	$-4 - 17$	(f)	$5 + (-4)$	(g)	$12 - (-7)$	(h)	$-8 - (+7)$
(i)	$20 - 31$	(j)	$6 - (-8)$	(k)	$-7 + (-7)$	(l)	$-12 - (-5)$
(m)	$38 + (-12)$	(n)	$-14 + (-3)$	(o)	$0 - 21$	(p)	$-24 - (-24)$
(q)	$9 - (5)$	(r)	$16 - (-16)$				

2. In the addition table, what is the value of $(m + n)$?

+	x	y	z
4	1	−5	m
e	7	n	10
f	2	−4	5

Note: Each number inside the table is the *sum* of its row and column number (e.g. $4 + x = 1$).

3. Simplify. (Do these calculations mentally, if possible!)

(a) 10×14

(b) $\frac{(-18)}{3}$

(c) $\frac{27}{(-9)}$

(d) $(-11) \times 0$

(e) $(-7) \times (-12)$

(f) $5 \times (-4)$

(g) $12 \times (-7)$

(h) $\frac{(-8)}{(-8)}$

(i) $20 \times \frac{30}{5}$

(j) $(-6) \times 8$

(k) $-9 \times \frac{(-12)}{21}$

(l) $\frac{(-12)}{(-3)} \times (-5)$

(m) $\frac{(-64)}{8} \times (-12)$

(n) $\frac{36}{(-9)} \times (-3)$

(o) $\frac{0}{(-21)}$

(p) $\frac{(-24)}{6} \times \frac{-16}{2}$

(q) $9 \times (-5) \times (-4)$

(r) $16 \times \frac{(-16)}{(-8)}$

4. In the multiplication table, what is the value of $(m - n)$?

×	2	c	d
a	6	−12	15
b	n	8	m

Note: Each number inside the table is the *product* of its row and column number (e.g. $a \times 2 = 6$).

5. Simplify. (Do these calculations mentally, if possible!)

(a) $6 \times 3 + 2(4)$

(b) $4 \times 6 - 3 \times 5$

(c) $\frac{18}{2} + 6 \times 5$

(d) $7 - \frac{12}{3} + 8$

(e) $15 + 3 \times 4 - 5$

(f) $9 + 9 \times 9 - 9$

(g) $20 \times 20 - \frac{306}{(-9)}$

(h) $28 - 7 \times 4 + 7$

(i) $\frac{42}{7} + 2 \times 10 - 10$

(j) $\frac{(18 + 6)}{(-6)} + 4$

(k) $7 - (8 + 2) \times \frac{4}{5}$

(l) $\frac{(8 + 3 \times 4)}{2} - 3 \times 4$

(m) $\frac{24}{(12 \times 2)} - \frac{10}{2}$

(n) $[(10 - 7) \times 3 + 5] \times 3 - 7$

(o) $(50 - 9 \times 5) \times 2 - \frac{40}{8}$

(p) $\frac{(19 - 7)}{3} - 4 \times (4 - 8)$

(q) $-7 \times 4 - \frac{16}{4} + 35$

(r) $8 - \left[2 + 3 \times \frac{(9 - 5)}{2}\right]$

4. Arithmetic terms

Product: A **product** is the answer (number) found after multiplication,
e.g. 32 is the product of 8 times 4, written: $8 \times 4 = 32$ (the product).

Quotient: A **quotient** is the answer (number) found after division,
e.g. 9 is the *quotient* after dividing 36 by 4, written: $\frac{36}{4} = 9$ (the quotient).

Divisor: A **divisor** is the number that is divided into another number to obtain the quotient. It is the *denominator* of a fraction,
e.g. for the division: $\frac{36}{9} = 4$; 9 is the *divisor* and 4 is the *quotient*.

Factor: A **factor** of a number (also known as a divisor) is an *integer* that divides into the number *without remainder*,
e.g. if the number is 8, then 1, 2, 4 and 8 are all factors of 8.
If the number is 21, then 1, 3, 7 and 21 are all factors of 21.

Alternatively, the *factors* of a number are any integer numbers which, when multiplied together, will equal that number,
e.g. the factors of 18 are (3 and 6), or (2 and 9), or (2, 3 and 3).
The factors of 84 are (4 and 21), or (3, 4 and 7), or (2, 2, 3 and 7).

Note: A factor is always *smaller than or equal to* the number of which it is a factor.

Multiple: For a given integer, a **multiple** of this number is any whole number that is exactly divisible by this number (i.e. without remainder).

Alternatively, a multiple of an integer number is the product found by multiplying this number by any natural number (i.e. by 1, 2, 3, 4, etc.),
e.g. for the number 4, multiples of 4 are 4, 8, 12, 16, 20, ... (since 4 can divide into each multiple without remainder);

e.g. for the number 7, multiples of 7 are 7, 14, 21, 28, ... (alternative explanation, since $7 \times 1 = 7$; $7 \times 2 = 14$; $7 \times 3 = 21$; $7 \times 4 = 28$; etc.).

Note: A multiple is always *equal to or larger* than the number of which it is a multiple.

Prime numbers: A **prime number** is any number that is divisible by itself and 1 only.
This means that a prime number *always* has only *two factors*, 1 and itself,
e.g. 2, 3, 5, 7, 11, 13, 17, 19, 23, 29, 31, 37 are the first 12 prime numbers.

Note 1 1 is not a prime number (as it has only one factor, namely 1).

Note 2 A negative number is never a prime number as it is divisible by itself, +1 and –1. Hence prime numbers are always only positive numbers.

Note 3 There is only one *even prime number* – the number 2.
Every other even number will have 2 as a factor – hence it cannot be a prime number.

Prime factors: Prime factors are the set of factors for a number that are only prime numbers.
To find prime factors, always reduce a number into its smallest factors, which must only be prime numbers.

Example 1	Example 2
Prime factors of 600	Prime factors of 252
Worked solution	**Worked solution**

Example 1
Prime factors of 600

Worked solution

```
2 | 600
2 | 300
2 | 150
3 |  75
5 |  25
5 |   5
  |   1
```

Answer: $2^3 \times 3 \times 5^2$

Example 2
Prime factors of 252

Worked solution

```
2 | 252
2 | 126
3 |  63
3 |  21
7 |   7
  |   1
```

Answer: $2^2 \times 3^2 \times 7$

Exercise 6

1. What is the product of:

    (a) 7 and 12   (b) 9 and –6   (c) 15 and 9   (d) 6 and 4 and 3

2. For each of the following, find the quotient and write down the divisor:

    (a) $\frac{(-12)}{4} = ?$ (b) $\frac{48}{(-12)} = ?$ (c) $\frac{63}{7} = ?$ (d) $\frac{27}{3} = ?$

3. Find all the factors of:

    (a) 24   (b) 10   (c) 17   (d) 28

    (e) 60   (f) 72   (g) 90   (h) 110

    (i) 50   (j) 51   (k) 37   (l) 84

4. Find the first 5 multiples of the following numbers:

    (a) 5   (b) 9   (c) 12   (d) 16

    (e) 20   (f) 50   (g) 250   (h) 600

5. List all the factors of the following numbers and state which are prime numbers:

    (a) 33   (b) 39   (c) 18   (d) 27

    (e) 31   (f) 51   (g) 53   (h) 43

    (i) 123   (j) 63   (k) 117   (l) 22

6. Write each number as the product of its prime factors:

    (a) 16   (b) 30   (c) 33   (d) 56

    (e) 19   (f) 45   (g) 78   (h) 132

    (i) 28   (j) 39   (k) 54   (l) 63

7. State whether true of false:

    (a) 28 is a prime number (b) 39 has only 3 factors (c) $\left(\frac{57}{3}\right)$ is a prime number

8. Which of these numbers cannot be expressed as the sum of two prime numbers?

    (a) 5   (b) 9   (c) 13   (d) 16   (e) 23

9. What is the greatest integer that is a sum of three different prime numbers, each less than 30?

10. If the product of the integers w, x, y and z is 770, and if $1 < w < x < y < z$, what is the value of $x + y$?

11. If x, y and z are positive integers such that x is a factor of y, and x is a multiple of z, and $z \neq x \neq y$, then which answers cannot be an integer?

    (a) $\frac{(x + z)}{z}$ (b) $\frac{y^2}{x}$ (c) $\frac{xy}{z}$ (d) $\frac{(y + z)}{x}$

12. If a positive integer, n, is divisible by both 3 and 5, then which of the following numbers will always be a divisor of n?

    (a) 8   (b) 15   (c) 30   (d) 60

13. If the quotient of $\frac{a}{b}$ is positive, then which of the following answers must be true?

    (a) $a < 0, b < 0$ (b) $a > 0, b > 0$ (c) $(a + b) > 0$ (d) $ab > 0$

14. How many multiples of 4 are there between 12 and 36 inclusive?

15. Which one of the following cannot be the greatest common divisor of two positive integers x and y?

    (a) 1   (b) x (c) y (d) $(x - y)$ (e) $(x + y)$

16. If p is the product of two prime numbers, p_1 and p_2, where $2 < p_1 < 6$ and $13 < p_2 < 20$, what is the smallest possible value of p?

17. If n is a prime number greater than 5, which one of the following terms must be even?

    (a) $(n + 2)^2$ (b) $n^2 + 2$ (c) $n(n + 2)$ (d) $(n + 1)(n + 2)$ (e) $(n - 2)^2$

18. x, y and z are positive integers such that $5 < x < y < z < 15$. x is even but not a multiple of 4; y is odd but not a prime number; and z is a prime number.

    What is the largest possible value of $x + y + z$?

5. Use of factors and multiples

a. Highest common factor (HCF)

The HCF is the largest factor that can divide into a set of numbers without remainder.
This is useful for cancelling common terms when performing arithmetic with fractions.

Example 1

Find the HCF of 24 and 40 (i.e. the largest factor common to both numbers).

Worked solution

Method 1

To find the HCF, first find the *prime factors* of each number.
$24 = \mathbf{2} \times \mathbf{2} \times \mathbf{2} \times 3$
$40 = \mathbf{2} \times \mathbf{2} \times \mathbf{2} \times 5$
Then find the product of the *common* prime factors.
HCF $= 2 \times 2 \times 2 = \mathbf{8}$

Method 2

Test whether the smallest number is the HCF. If it can divide into *all the numbers* without remainder, then it is the HCF.
If not, find all the factors – in decreasing size – of this smallest number. Test each factor, in decreasing order, for the HCF.

Choose 24 as the HCF of 24 and 40.
Since 24 does not divide into 40, find the factors – in decreasing order – of 24. They are 12, 8, 6, 4 and 2.
First test whether 12 can divide into both 24 and 40 – the answer is 'no'
Next test whether 8 can divide into both 24 and 40 – the answer is 'yes'.
Hence **8** is the HCF of 24 and 40.

b. **Lowest common multiple (LCM) or lowest common denominator (LCD)**

The LCM is the lowest number that all the numbers can divide into without remainder (i.e. the *smallest common multiple* of the numbers).
This is useful when adding and/or subtracting fractions.

Example 1

Find the LCM of 48 and 60.

Worked solution

Method 1

To find the LCM, first find the *prime factors* of each number.
$48 = \mathbf{2} \times \mathbf{2} \times \underline{2} \times \underline{2} \times \mathbf{3}$
$60 = \mathbf{2} \times \mathbf{2} \times \mathbf{3} \times \underline{5}$
Then find the product of the *common* prime factors (HCF) multiplied by the product of the *non-common* factors:
$LCM = (\mathbf{2} \times \mathbf{2} \times \mathbf{3}) \times (\underline{2} \times \underline{2} \times \underline{5}) = 240$

Example 2

Find the LCM of 9 and 15.

Worked solution

Method 2

Test whether the largest number is the LCM. If all the numbers can divide into this number without remainder, then it is the LCM.
If not, choose consecutive multiples of the largest number and test whether all the other numbers can divide into this multiple without remainder.

Find the LCM of 9 and 15.
Choose 15 as the LCM of 9 and 15.
Since 9 does not divide into 15, take consecutive multiples of 15.
Multiples of 15 are 15, 30, 45, 60, 75, etc.
Test whether 9 and 15 can both divide into 30 – the answer is 'no'
Test whether 9 and 15 can both divide into 45 – the answer is 'yes'.
Hence **45** is the LCM of 9 and 15.

Exercise 7

1. Find the HCF of the following numbers:
   (a) 12 and 30
   (b) 56 and 42
   (c) 63 and 84
   (d) 75 and 105
   (e) 36, 44 and 68
   (f) 28 and 98
   (g) 33 and 132
   (h) 36 and 90
   (i) 125 and 300
   (j) 64, 112 and 256
   (k) 18, 36 and 72

2. Find the LCM of the following numbers:
   (a) 8 and 20
   (b) 12 and 16
   (c) 6, 9 and 12
   (d) 15, 18 and 30
   (e) 14, 15 and 20
   (f) 6 and 15
   (g) 15 and 21
   (h) 10 and 16
   (i) 5, 10 and 15
   (j) 14, 21 and 35

3. State whether true or false:
   (a) The LCM of 9 and 8 is 72.
   (b) The LCM of 8 and 18 is 144.
   (c) The HCF of 26, 52 and 104 is 13.
   (d) The HCF of 48, 64 and 144 is 8.

4. If a positive integer is divisible by 5 and 8, then this integer must also be divisible by which of the following?
   (a) 13
   (b) 20
   (c) 40
   (d) 80

1.4 Fractions

A fraction is a part of a whole number $\left(\text{e.g. } \frac{3}{5}, -\frac{4}{7}, \frac{9}{16}, \frac{3}{4}, -\frac{8}{11}, \frac{1}{6}\right)$.

1. Types of fractions

A fraction number can be written as either:

- a **mixed fraction**, which has an integer part and a fraction part $\left(\text{e.g. } 4\frac{1}{2}, 7\frac{2}{3}, 19\frac{1}{4}\right)$, or as
- an **improper fraction**, where the integer part of a mixed fraction is included in the numerator and the fraction number is then written entirely as a fraction $\left(\text{e.g. } \frac{14}{3}, -\frac{5}{4}, \frac{24}{5}\right)$.

Note: A mixed fraction can be written as an improper fraction by multiplying the integer part of the number by the denominator of the fraction and then adding the numerator value to this product.

To do arithmetic with fractions, a mixed fraction number must always be converted to an improper fraction.

Example 1	**Example 2**
Convert the mixed fraction $4\frac{2}{3}$ to an improper fraction.	Convert the mixed fraction $12\frac{3}{4}$ to an improper fraction.
Worked solution	**Worked solution**
$4\frac{2}{3} = \frac{(4 \times 3 + 2)}{3} = \frac{14}{3}$	$12\frac{3}{4} = \frac{(12 \times 4 + 3)}{4} = \frac{51}{4}$

---

2. Arithmetic operations with fractions

a. Adding and subtracting fractions

e.g. $3\frac{1}{4} \pm 2\frac{2}{3}$

1. Write each fraction as an improper fraction.
2. Find the LCM of the denominators.
3. Divide each fraction's denominator into the LCM and multiply the quotient by the fraction's numerator. This converts (or scales) each fraction into the same LCM denominator.
4. Add and/or subtract the (scaled) numerators.
5. Simplify the resultant fraction by cancelling common terms.

Example 1	**Example 2**
Solve $\frac{3}{5} - \frac{1}{6} + \frac{2}{3}$.	Solve $\frac{3}{4} - \frac{1}{6}$.
Worked solution	**Worked solution**
The LCM of 3, 5 and 6 is 30. (Each divides into 30 without remainder.)	The LCM of 4 and 6 is 12.
$\frac{3(6) - 1(5) + 2(10)}{30} = \frac{18 - 5 + 20}{30} = \frac{33}{30} = \frac{11}{10}$	$\frac{3(3) - 1(2)}{12} = \frac{9 - 2}{12} = \frac{7}{12}$
$= 1\frac{1}{10}$	

b. Multiplying fractions

e.g. $6\frac{2}{5} \times 4\frac{3}{8}$

1. Write all numbers as improper fractions.
2. Where possible, cancel common terms across the numerators and denominators.
3. Then multiply numerators together and denominators together.
4. Write the answer as a mixed fraction, if appropriate.

Example 1	**Example 2**
Calculate $\frac{5}{9} \times \frac{3}{4} \times \frac{8}{15}$.	Calculate $3\frac{2}{3} \times 2\frac{1}{4}$.
Worked solution	**Worked solution**
$\frac{5}{9} \times \frac{3}{4} \times \frac{8}{15}$ (cancel 3, 4 and 5)	$3\frac{2}{3} \times 2\frac{1}{4} = \frac{11}{3} \times \frac{9}{4}$ (cancel 3)
$= \frac{1 \times 1 \times 2}{3 \times 1 \times 3} = \frac{2}{9}$	$= \frac{11}{1} \times \frac{3}{4}$
	$= \frac{33}{4} = 8\frac{1}{4}$

c. Dividing fractions

e.g. $3\frac{1}{8} \div \frac{5}{6}$

1. Write both the numerator and denominator fractions as improper fractions.
2. Now invert the denominator fraction and multiply the two fractions together. Follow the method for multiplying fractions (see above).
3. Write the answer (the product of fractions) as a mixed fraction.

Example 1	Example 2
Calculate $\frac{7}{12} \div \frac{2}{9}$.	Calculate $\frac{4\frac{3}{4}}{1\frac{2}{3}}$.
Worked solution	**Worked solution**
$\frac{7}{12} \div \frac{2}{9} = \frac{7}{12} \times \frac{9}{2}$ (cancel 3)	$\frac{4\frac{3}{4}}{1\frac{2}{3}} = \frac{\frac{19}{4}}{\frac{5}{3}} = \frac{19}{4} \times \frac{3}{5}$
$= \frac{7}{4} \times \frac{3}{2}$	$= \frac{57}{20} = 2\frac{17}{20}$
$= \frac{21}{8} = 2\frac{5}{8}$	

Exercise 8

1. Simplify (write the answer as a mixed fraction, if appropriate):

   (a) $\frac{7}{8} - \frac{2}{3}$ (b) $\frac{5}{9} + \frac{1}{4}$ (c) $\frac{6}{7} + \frac{2}{3}$ (d) $3\frac{1}{4} + 1\frac{1}{5}$ (e) $\frac{2}{3} - 2\frac{1}{8}$

2. Simplify (write the answer as a mixed fraction, if appropriate):

   (a) $\frac{7}{8} \times \frac{2}{3}$ (b) $\frac{5}{9} \times 1\frac{4}{5}$ (c) $\frac{6}{7} \times \frac{2}{3}$ (d) $3\frac{1}{4} \times 1\frac{1}{5}$ (e) $\frac{2}{3} \times 2\frac{1}{8}$

3. Simplify (write the answer as a mixed fraction, if appropriate):

   (a) $\frac{7}{8} \div \frac{2}{3}$ (b) $\frac{5}{9} \div 1\frac{4}{5}$ (c) $\frac{6}{7} \div \frac{2}{3}$ (d) $3\frac{1}{4} \div 1\frac{1}{5}$ (e) $\frac{2}{3} \div 2\frac{1}{8}$

4. Simplify:

   (a) $\left(\frac{3}{4} - \frac{2}{3}\right) \div \left(\frac{3}{4} + \frac{2}{3}\right)$ (b) $\left(2\frac{1}{4} \times \frac{2}{3}\right) \div \left(\frac{3}{5} - 1\frac{1}{4}\right)$ (c) $\left(3\frac{1}{2} - 2\frac{2}{3}\right) \div \left(\frac{1}{2} \times 1\frac{3}{8}\right)$

   (d) $\left(\frac{5}{6} + \frac{1}{3}\right) \div \left(1 - \frac{5}{6}\right)$ (e) $\left(3\frac{3}{8} - \frac{3}{4}\right) \times \left(1\frac{2}{3} + 1\frac{4}{9}\right)$ (f) $\left(1\frac{2}{3} + \frac{1}{5}\right) \times \left(2\frac{1}{4} - 1\frac{2}{3}\right)$

5. There are n people in a room. If $\frac{3}{7}$ of these people are under 21 years, and $\frac{5}{13}$ are older than 65 years, how many people in the room are between 21 and 65 years, where $50 < n < 100$?

1.5   Decimals

The fraction part of a number can be written either as a *fraction* $\left(\frac{1}{4}, \frac{1}{2}, \frac{5}{8}\right)$ or as a *decimal* (0.25 ; 0.5 ; 0.625).

A **decimal** can be defined as a fraction with an implied denominator of either:

10 if the number has only one decimal place $\left(\text{e.g. } 0.4 = \frac{4}{10}\right)$

100 if the number has only two decimal places $\left(\text{e.g. } 0.64 = \frac{64}{100}\right)$

1 000 if the number has only three decimal places $\left(\text{e.g. } 0.218 = \frac{218}{1\,000}\right)$, etc.

Exercise 9

1. Arrange the numbers in order of size, starting with the smallest.
   - (a) 0.04, 0.4, 0.38
   - (b) 0.809, 0.81, 0.8
   - (c) 1, 0.99, 0.998, 0.0999
   - (d) 4.03, 3.99, 4.0, 4.024

2. True or false?
   - (a) $0.078 < 0.08$
   - (b) $0.88 < 0.089$
   - (c) $0.001 > 0.0001$
   - (d) $3.65 > 3.605$
   - (e) $0.31 > 0.3100$

3. Which of the following is greater than $\frac{2}{3}$?
   - (a) $\frac{16}{25}$
   - (b) $\frac{8}{12}$
   - (c) $\frac{3}{5}$
   - (d) $\frac{8}{11}$
   - (e) $\frac{5}{8}$

4. Convert to decimals, and then find the value of $\left(\frac{1}{5}\right)^2 - (0.2)\left(\frac{1}{4}\right)$.

1. Multiplication and division by 10, 100, 1 000, etc.

If a number is *multiplied* by 10 or 100 or 1 000, move the decimal point to the *right* as many times as there are zeros in the *tens* number.

Example 1	**Example 2**
Calculate 45.24 × 10.	Calculate 3.745 × 100.
Worked solution	**Worked solution**
45.24 × 10 = 452.4	3.745 × 100 = 374.5

Example 3	**Example 4**
Calculate 0.1189 × 1 000.	Calculate 19.784 × 10 000.
Worked solution	**Worked solution**
0.1189 × 1 000 = 118.9	19.784 × 10 000 = 197 840

If a number is *divided* by 10 or 100 or 1 000, move the decimal point to the *left* as many times as there are zeros in the *tens* number.

Example 1	**Example 2**
Calculate 45.24 ÷ 10.	Calculate 3.745 ÷ 100.
Worked solution	**Worked solution**
45.24 ÷ 10 = 4.524	3.745 ÷ 100 = 0.03745

Example 3	**Example 4**
Calculate 0.1189 ÷ 1 000.	Calculate 19.784 ÷ 10 000.
Worked solution	**Worked solution**
0.1189 ÷ 1 000 = 0.0001189	19.784 ÷ 10 000 = 0.0019784

2. Changing a fraction to a decimal

To convert a fraction to a decimal:
1. Place a decimal point after the numerator number and add zeros.
2. Divide the denominator into the numerator number until there is no remainder (or until the division is recurring).

Example 1

Convert $\frac{5}{8}$ to a decimal.

Worked solution

$\frac{5}{8} = \frac{5.000}{8} = 0.625$

Example 2

Convert $\frac{6}{11}$ to a decimal.

Worked solution

$\frac{6}{11} = \frac{6.0000}{11} = 0.5454$ (recurring)

Example 3

Convert $3\frac{7}{8}$ to a decimal.

Worked solution

$3\frac{7}{8} = 3 + \frac{7.0000}{8} = 3 + 0.875 = 3.875$

3. Addition and subtraction with decimal numbers

To add or subtract decimal numbers:
1. Line up the digits of the numbers, one under the other, on the decimal point.
2. Once the numbers have been aligned on the decimal point, fill in zeros to make the lengths of the decimal part of the numbers the same.
3. Add (or subtract) in the normal way (i.e. as with whole numbers).

Example 1

Calculate 7.48 + 2.6433.

Worked solution

```
   7.4800
+  2.6433
 10.1233
```

Example 2

Calculate 8.3 – 5.0628.

Worked solution

```
   8.3000
−  5.0628
   3.2372
```

Example 3

Calculate 15.432 – 11.04.

Worked solution

```
  15.432
− 11.040
   4.392
```

Exercise 10

1. Calculate:
   - (a) 0.064×100
   - (b) $72.75 \times 1\,000$
   - (c) 8.125×100
   - (d) 0.0045×10
   - (e) 7.877×100
   - (f) $3.6 \div 100$
   - (g) $19.784 \div 10$
   - (h) $0.0082 \div 100$
   - (i) $12\,723.94 \div 1\,000$
   - (j) $80.003 \div 1\,000$

2. Convert the following fractions to decimals:
   - (a) $\frac{3}{8}$
   - (b) $\frac{5}{9}$
   - (c) $\frac{3}{10}$
   - (d) $\frac{5}{12}$
   - (e) $\frac{5}{13}$
   - (f) $\frac{5}{6}$
   - (g) $\frac{3}{50}$
   - (h) $\frac{7}{16}$

3. Solve the following (without using a calculator):
   - (a) $25.96 + 1.754$
   - (b) $12.05 + 9.265$
   - (c) $8.125 + 12$
   - (d) $0.0045 + 0.058$
   - (e) $7.77 + 77.7$
   - (f) $3.6 - 2.24$
   - (g) $19.784 - 0.88$
   - (h) $7 - 0.035$
   - (i) $23.94 - 8.004$
   - (j) $77.7 - 7.707$

4. Find the decimal value of: $\frac{7}{100} + \frac{3}{1\,000} + \frac{9}{100\,000}$

4. Multiplication with decimal numbers

To multiply with decimal numbers:
1. Ignore the decimal points in the numbers (i.e. treat them all as integers).
2. Multiply the integer numbers in the usual arithmetic way to obtain the product.
3. Count the number of decimal places in total in the original numbers.
4. Move the decimal point in the product (found in 2) to the left by the total number of decimal places in the original decimal numbers (found in 3).

Example 1

Solve 4.1×1.2.

Worked solution

Ignore decimal places.
$41 \times 12 = 492.0$

In total: two decimal places
Now move two decimal places to the left from $492.0 = 4.92$

Example 2

Solve 0.82×0.07.

Worked solution

Ignore decimal places.
$82 \times 7 = 574.0$

In total: four decimal places
Now move four decimal places to the left from $574.0 = 0.0574$

Example 3

Solve 0.014×3.04.

Worked solution

Ignore decimal places.
$14 \times 304 = 4\,256.0$

In total: five decimal places
Now move five decimal places to the left from 4 256.0 = 0.04256

5. Division with decimal numbers

To divide with decimal numbers:
1. Convert *only the divisor* to a whole number (integer) by multiplying it by its implied denominator of 10, 100, 1 000, etc.
2. Multiply the number being divided by the same multiple of 10 (to preserve the ratios of the numbers to each other).
3. Divide the (modified) numbers into each other in the normal arithmetic way.

Example 1

Solve 2.76 ÷ 0.3.

Worked solution

2.76 ÷ 0.3
Multiply both numbers by 10
27.6 ÷ 3 = 9.2

Example 2

Solve 0.826 ÷ 0.07.

Worked solution

0.826 ÷ 0.07
Multiply both numbers by 100
82.6 ÷ 7 = 11.8

Example 3

Solve 6.231 ÷ 3.1.

Worked solution

6.231 ÷ 3.1
Multiply both numbers by 10: 62.31 ÷ 31 = 2.01

Exercise 11

1. Solve (without using a calculator):

    (a) 6.5 × 0.8   (b) 5.2 × 0.006   (c) 1.33 × 0.04

    (d) 2.04 × 0.0004   (e) 3.003 × 0.01   (f) 0.83 × 1.5

    (g) 5 × 4.106   (h) 7.92 × 0.04   (i) 0.002 × 4.16

    (j) 77.7 × 0.003

2. Solve (without using a calculator) to four decimal places, if necessary:

    (a) $\frac{62.4}{0.8}$ (b) $\frac{5.22}{0.6}$ (c) $\frac{1.233}{0.03}$

    (d) 2.04 ÷ 0.0004   (e) 3.003 ÷ 0.01   (f) $\frac{0.0825}{1.5}$

    (g) 41.05 ÷ 0.5   (h) $\frac{7.08}{0.04}$ (i) $\frac{0.04414}{0.002}$

    (j) 77.7 ÷ 0.003

3. Solve (without using a calculator):

    (a) 0.838 × 10   (b) 0.074 × 100   (c) 0.6105 × 100

    (d) 0.00045 × 1 000   (e) 23.6 ÷ 1 000   (f) 0.72 ÷ 10

    (g) 2.111 ÷ 100   (h) 0.085 ÷ 100   (i) 23.5864 × 10 ÷ 1 000

4. Solve (without using a calculator):

(a)	3.7 + 0.65	(b)	11.3 – 2.14	(c)	2.52 × 0.4
(d)	17.1 + 3.55	(e)	12 – 1.8	(f)	23.6 ÷ 8
(g)	12.4 × 0.05	(h)	0.028 × 100	(i)	18.606 ÷ 7
(j)	34.64 × 1 000	(k)	6.04 × 11	(l)	0.048 ÷ 10
(m)	8.2 – 1.55	(n)	58.4 ÷ 10 000	(o)	0.072 ÷ 8
(p)	7.05 – 3.6	(q)	(18 + 1.8 + 0.18) ÷ 1 000		
(r)	33.3 × 0.007	(s)	0.392 ÷ 0.04	(t)	586.26 ÷ 9

1.6   Indices (or powers)

A **power** means that the **base number** is multiplied by itself as many times as shown by the power, i.e. $a^x = a \times a \times a \times ... \times a$

Example 1	Example 2
$3^4 = 3 \times 3 \times 3 \times 3 = 81$	$5^3 = 5 \times 5 \times 5 = 125$

1. Base^0

Any base number (integer or fraction) raised to the *power of zero* (0) is **always 1** (not zero!).

Example 1	Example 2
$9^0 = 1$	$(-62)^0 = 1$

Example 3	Example 4
$\left(\frac{3}{4}\right)^0 = 1$	$\left(4\frac{3}{5}\right)^0 = 1$

2. Negative powers

When a base number and its power is moved across the division line, the sign of the power changes, i.e. $a^{-x} = \frac{1}{a^x}$

Example 1	Example 2
$3^{-4} = \frac{1}{3^4} = \frac{1}{81}$	$5^{-3} = \frac{1}{5^3} = \frac{1}{125}$

3. Multiplying with powers

To simplify, *add* the powers of like bases, i.e. $a^x \times a^y = a^{(x+y)}$

Example 1	Example 2
Simplify $3^4 \times 3^2$.	Simplify $7^2 \times 7^3$.
Worked solution	**Worked solution**
$3^4 \times 3^2 = 3^{(4+2)} = 3^6$	$7^2 \times 7^3 = 7^{(2+3)} = 7^5$

4. Dividing with powers

To simplify, *subtract* the powers of like bases, i.e. $a^x \div a^y = a^{(x-y)}$

Example 1

Simplify $3^5 \div 3^2$

Worked solution

$3^5 \div 3^2 = 3^{(5-2)} = 3^3 = 27$

Example 2

Simplify $7^2 \div 7^3$

Worked solution

$7^2 \div 7^3 = 7^{(2-3)} = 7^{-1} = \frac{1}{7}$

5. Raising powers to powers

When a power is raised to a power, the powers are *multiplied* together,
i.e. $(a^x)^y = a^{xy}$

Example 1

Simplify $(3^4)^2$

Worked solution

$(3^4)^2 = 3^{(4 \times 2)} = 3^8$

Example 2

Simplify $(7^2)^3$

Worked solution

$(7^2)^3 = 7^{(2 \times 3)} = 7^6$

6. Distributing powers

When a base is factorised (or is a fraction), each factor (or term in the fraction) is given the same power, i.e. $(ab)^x = a^x \times b^x$ and $\left(\frac{a}{b}\right)^x = \frac{a^x}{b^x}$

Example 1

Simplify 10^3

Worked solution

$10^3 = (2 \times 5)^3 = 2^3 \times 5^3$

Example 2

Simplify 12^4

Worked solution

$12^4 = (2 \times 2 \times 3)^4 = (2^2 \times 3)^4 = 2^8 \times 3^4$

Example 3

Simplify $\left(\frac{3}{5}\right)^2$

Worked solution

$\left(\frac{3}{5}\right)^2 = \frac{3^2}{5^2}$

Example 4

Simplify $\left(2\frac{2}{3}\right)^3$

Worked solution

$\left(2\frac{2}{3}\right)^3 = \left(\frac{8}{3}\right)^3 = \frac{8^3}{3^3}$

7. Grouping powers

When different bases (factors) have the *same power*, the bases can be multiplied (or divided) together and given the same power,
i.e. $a^x \times b^x = (ab)^x$ and $= \frac{a^x}{b^x} = \left(\frac{a}{b}\right)^x$

Example 1

Simplify $3^2 \times 5^2$

Example 2

Simplify $\frac{15^4}{3^4}$

Worked solution	Worked solution
$3^2 \times 5^2 = (3 \times 5)^2 = 15^2 = 225$	$\frac{15^4}{3^4} = \left(\frac{15}{3}\right)^4 = 5^4 = 625$

Exercise 12

1. Solve the following (without using a calculator):

   (a) 3^3 (b) 1^8 (c) 6^3 (d) 10^4

   (e) 2^6 (f) $(-4)^3$ (g) $(-1)^9$ (h) $(-5)^4$

   (i) $(-10)^4$ (j) 8^0

2. Solve the following (without using a calculator):

   (a) 3^{-3} (b) 1^{-6} (c) $(-4)^{-4}$ (d) 10^{-4}

   (e) 2^{-10} (f) $(-1)^{-17}$ (g) $(-10)^{-3}$ (h) 5^{-2}

   (i) 100^{-3}

3. Simplify the following and write the answer as a real number:

   (a) $6^4 \times 6^5$ (b) 8×8^3 (c) $3^{-4} \times 3^3$ (d) $5^{-3} \times 5^3$

   (e) $10^7 \times 10^2$ (f) $1^4 \times 1^{-7}$ (g) $4^0 \times 4^3$ (h) $7^7 \times 7^{-3}$

   (i) $9^0 \times 9^{-2}$ (j) $2^{-3} \times 2^{-2}$ (k) $9^2 \times 3^{-2}$

   (l) $6^{-2} \times 6^5 \div 6^2$ (m) $5^2 \times 25 \div 5^5$

4. Simplify the following and write the answer as a real number:

   (a) $6^4 \div 6^5$ (b) $8 \div 8^3$ (c) $3^{-4} \div 3^3$ (d) $5^{-3} \div 5^{-2}$

   (e) $10^7 \div 10^2$ (f) $1^4 \div 1^{-7}$ (g) $8^4 \div 8^4$ (h) $8^{-1} \div 8^4$

   (i) $5^1 \div 5^{-3}$ (j) $10^{-3} \div 10^{-7}$ (k) $2^3 \times 2^0 \div 2^2$

   (l) $16 \times 2^2 \div 8^2$ (m) $36^2 \div 6^3 \times 6^{-2}$

5. Simplify the following and write the answer as a real number:

   (a) $(2^4)^5$ (b) $(8^3)^{(-2)}$ (c) $(3^{-2})^2$ (d) $(5^{-3})^{(-2)}$

   (e) $(10^5)^2$ (f) $(1^4)^{(-7)}$ (g) $(7^4)^0$

6. Answer true or false:

   (a) $2^3 = 8$ (b) $3^2 = 6$ (c) $2^{-1} = \frac{1}{2}$ (d) $10^{-2} = \frac{1}{20}$

   (e) $2^3 > 3^2$ (f) $1^9 = 9$ (g) $(-3)^2 = -9$ (h) $10^{-1} = 0.1$

   (i) $2^{-2} = 0.25$ (j) $2^4 = 4^2$ (k) $2^5 = 5^2$ (l) $(-2)^3 = 2^3$

   (m) $1^{10} = \frac{1}{10}$ (n) $3^4 < 4^3$ (o) $2^{-3} > 1$ (p) $7^2 > 50^1$

   (q) $6^0 = 6$ (r) $8^5 \div 2^{10} = 32$ (s) $9^2 \times 3^0 = 3^4$ (t) $5^{-3} < 0.125$

7. Factorise the bases of the following numbers and leave the answer in index (powers) form:

   (a) 15^3 (b) 22^2 (c) -6^3 (d) 39^2

8. Express the following numbers with a single base:

   (a) $4^3 \times 5^3$ (b) $7^2 \times 3^2$ (c) $\frac{27^2}{9^2}$ (d) $\frac{(-18)^3}{-6^3}$

(e) $2^4 \times 5^2$ (f) $2^4 \times 3^4 \times 5^4$ (g) $\frac{4^3 \times 7^3}{2^3}$

9. What is the value of $(m - k)$ if $\frac{(0.0015 \times 10^m)}{(0.03 \times 10^k)} = 5 \times 10^7$?

10. Solve for the value of $2 + 2 + 2^2 + 2^3 + 2^4$. Use factors and index laws.

11. Simplify $(2^3 + 2^3) \times (3^3 + 3^3 + 3^3)$. Leave the answer in index form.

1.7   Percentages

A **percentage** is a number that represents a **fraction out of 100**,

e.g. 45% means $\frac{45}{100} = \frac{9}{20}$

The following covers basic operations with percentages.

1. Convert fractions to percentages

1. Multiply the numerator of the fraction by 100.
2. Divide by the denominator.

Example 1

Convert $\frac{3}{8}$ to a percentage.

Worked solution

$\frac{3}{8} = (3 \times 100) \div 8 = \frac{300}{8} = 37.5\%$

Example 2

Convert $\frac{2}{9}$ to a percentage.

Worked solution

$\frac{2}{9} = (2 \times 100) \div 9 = \frac{200}{9} = 22.22\%$

Example 3

What percentage is 7 out of 20?

Worked solution

As a fraction it is $\frac{7}{20}$
Convert to % = $(7 \times 100) \div 20$
$= \frac{700}{20} = 35\%$

Example 4

What percentage is R90 of R400?

Worked solution

As a fraction it is $\frac{90}{400}$ or $\frac{9}{40}$
Convert to % = $(9 \times 100) \div 40$
$= \frac{900}{40} = 22.5\%$

2. Percentage of a number

1. Write the percentage as a fraction out of 100.
2. Multiply the number by this percentage as a fraction and simplify.

Example 1

What is 20% of R75?

Worked solution

Solve $\frac{20}{100} \times 75$

$= \frac{1}{5} \times 75 = R15$

Example 2

What is 45% of 120 kg?

Worked solution

Solve $\frac{45}{100} \times 120$

$= \frac{9}{20} \times 120 = 54$ kg

3. Percentage increases and decreases

$$\text{Percentage increase} = \frac{\text{actual increase}}{\text{initial (base) number}} \times 100$$

$$\text{Percentage decrease} = \frac{\text{actual decrease}}{\text{initial (base) number}} \times 100$$

Note: Always use the *original (initial or start) value* as the denominator in percentage increase/decrease calculations.

Example 1

Calculate the *percentage increase* from R75 to R90.

Worked solution

Increase = R15
Start value = R75

Then $\frac{15}{75} \times 100 = 20\%$

Example 2

Calculate the *percentage increase* from R21 to R28.35.

Worked solution

Increase = R7.35
Start value = R21

Then $\frac{7.35}{21} \times 100 = 35\%$

Example 3

Calculate the *percentage decrease* from R120 to R105.

Worked solution

Decrease = R15
Start value = R120

Then $\frac{15}{120} \times 100 = 12.5\%$

Example 4

Calculate the *percentage decrease* from R7.50 to R6.00.

Worked solution

Decrease = R1.50
Start value = R7.50

Then $\frac{1.5}{7.5} \times 100 = 20\%$

4. Increase/decrease numbers by a percentage

Method 1 Find the percentage increase (or decrease) and add it to (or subtract it from) the original number.

Method 2 The final answer after the % increase (or % decrease) can be found directly as follows:

- Number after % increase = $\frac{(100 + \% \text{ increase})}{100} \times$ original number

- Number after % decrease = $\frac{(100 - \% \text{ decrease})}{100} \times$ original number

Example 1
Increase 16 by 25%.

Example 2
Decrease 16 by 25%.

Worked solution	Worked solution
Method 1	**Method 1**
25% of 16 = 4	25% of 16 = 4
Then answer = 16 + 4 = 20	Then answer = 16 − 4 = 12
Method 2	**Method 2**
Answer = $\frac{(100 + 25)}{100} \times 16$	Answer = $\frac{(100 - 25)}{100} \times 16$
$= \frac{125}{100} \times 16 = 20$	$= \frac{75}{100} \times 16 = 12$

5. Find the total from a given part (% or fraction) of the whole

Note: Always let the whole (total number) = 100 (as a %) or 1 (as a fraction).

$$\text{Total (whole)} = \frac{100}{\text{(given \%)}} \times \text{value of given \%, or}$$

$$\text{Total (whole)} = \frac{1}{\text{(given fraction)}} \times \text{value of given fraction}$$

Example 1
If 5% of an amount is R12, what is the total amount?

Worked solution

Since 5% = R12, then the total amount = $\frac{100}{5} \times 12 = R240$

Example 2
If $\frac{1}{4}$ of an amount is R21, what is the total amount?

Worked solution

Since $\frac{1}{4}$ = R21, then the total amount = $\frac{1}{\left(\frac{1}{4}\right)} \times 21 = \frac{4}{1} \times 21 = R84$

Example 3
When a product is discounted by 10%, it sells for R45. What was its original selling price?

Worked solution

The initial (or base) number is the original selling price = 100%

The discounted selling price = (100 − 10) = 90%

Now, if 90% = R45, then 100% = $\frac{100}{90} \times 45 = R50$

Example 4
When a product is marked up by 25%, it sells for R65. What was its cost price?

Worked solution

The initial (or base) number is the cost price = 100%

The selling price (after a 25% mark-up) = (100 + 25) = 125%

Now, if 125% = R64, then 100% = $\frac{100}{125}$ × 65 = R52

Example 5

When an amount is increased by $\frac{1}{5}$, its value is R36. What was the initial amount?

Worked solution

Note that the increase is a fraction of 1. The initial (or base) amount = 1

The increased amount = $\left(1 + \frac{1}{5}\right) = \frac{6}{5}$ (in fraction terms)

Now, if $\frac{6}{5}$ = R36, then the initial amount (1) = $\frac{1}{\frac{6}{5}}$ × 36 = $\frac{5}{6}$ × 36 = R30

Example 6

When an amount is decreased by $\frac{1}{8}$, its value is R84. What was the initial amount?

Worked solution

Note that the decrease is a fraction of 1. The initial (or base) amount = 1

The decreased amount = $\left(1 - \frac{1}{8}\right) = \frac{7}{8}$ (in fraction terms)

Now, if $\frac{7}{8}$ = R84, then the initial amount (1) = $\frac{1}{\left(\frac{7}{8}\right)}$ × 84 = $\frac{8}{7}$ × 84 = R96

Exercise 13

1. Write each fraction as a percentage:

   (a) $\frac{7}{8}$ (b) $\frac{5}{9}$ (c) $\frac{5}{12}$ (d) $\frac{7}{16}$ (e) $\frac{4}{15}$

   (f) $\frac{1}{12}$ (g) $\frac{7}{20}$

2. What percentage is each of the following?

   (a) 14 out of 25   (b) R42 of R60   (c) 56 kg of 800 kg

   (d) 17 km of 40 km   (e) R55 of R250

3. Calculate:

   (a) 5% of 60   (b) 9% of R400   (c) 12% of 800 kg

   (d) 30% of 350 g   (e) 15% of R250

4. (a) In a car park, there are 35 white cars and 105 cars of other colours. What percentage of the cars is white?

   (b) John, Mick and Jane are the owners of a company. John owns 54 shares, Mick owns 81 shares and Jane owns 45 shares in the company. What percentage of the company does each person own?

(c) A street vendor who sells magazines receives 15% of all sales as his income. If he sells 150 magazines at R20 each, what is his income?

(d) An insurance broker receives 4% commission on each policy she sells. If she sells 5 policies, each worth R10 000, how much does she earn in commission?

(e) An estate agent receives 5% on the first R100 000 and 2% on the balance on the sale of a property. How much commission does he receive on a property sold for R380 000?

5. Calculate the percentage increase:

    (a) from R80 to R90   (b) from 75 kg to 117 kg

    (c) from R240 to R300   (d) from 1 100 km to 1 210 km

6. (a) If a product is bought for R21 and sold for R28, what is the percentage mark-up?

    (b) If a product is bought for R112 and sold for R154, what is the percentage mark-up?

    (c) Before a run, an athlete's heart rate was 60 beats per minute. After the run, her heart rate was 75 beats per minute. What was the percentage increase in her heart rate?

7. Calculate the percentage decrease:

    (a) from R105 to R84   (b) from 120 kg to 105 kg

    (c) from R880 to R836   (d) from 45 km to 30 km

8. (a) If a product is originally marked as R64 but sold for R48, what is the percentage loss?

    (b) A person weighed 120 kg before a diet and 84 kg afterwards. What is the percentage reduction in weight?

    (c) At the start of the season an athelete's heart rate was 80 beats per minute. At the end of the season, it was 64 beats per minute. By what percentage did his heart rate decrease during the season?

    (d) In 2006, the book value of a certain car was $\frac{2}{3}$ of the original purchase price, and in 2008 its book value was $\frac{1}{2}$ of the original purchase price. By what percentage did the book value of the car decrease from 2006 to 2008?

9. Find the answer to the following percentage increases:

    (a) R30 by 20%   (b) R120 by 15%   (c) 75 kg by 40%

    (d) 500 cm by 50%   (e) R70 by 5%

10. Find the answer to the following percentage decreases:

    (a) 40 kg by 20%   (b) R240 by 25%   (c) 75 cm by 40%

    (d) 15 cm by 30%   (e) R150 by $33\frac{1}{3}$%

11. Find the whole amount if:

    (a) 25% is equal to 14 kg   (b) 15% is equal to R30

    (c) 20% is equal to 120 km (d) 12% is equal to 36 g

(e) $\frac{1}{3}$ is equal to R17   (f) $\frac{1}{8}$ is equal to 9 kg

(g) $\frac{3}{8}$ is equal to R48   (h) $\frac{2}{5}$ is equal to 32 km

12. Find the initial amount if the final amount, after an increase of:
    (a) 20% was equal to R84   (b) 25% was equal to 150 kg
    (c) 10% was equal to 55 km   (d) 40% was equal to 560 g
    (e) $\frac{1}{5}$ was equal to 240 km   (f) $\frac{1}{8}$ was equal to R180

13. Find the initial amount if the final amount, after a decrease of:
    (a) 20% was equal to R72   (b) 25% was equal to 120 kg
    (c) 10% was equal to 45 km   (d) 40% was equal to 360 g
    (e) $\frac{1}{4}$ was equal to R300   (f) $\frac{1}{6}$ was equal to 80 m

14. A concert organiser predicted a 25% increase in attendance this year over that of last year, but actual attendance this year decreased by 20%. What percentage of the projected attendance was the actual attendance?

15. In a school, 60% of learners weigh less than 75 kg. A total of 48 learners weigh less than 50 kg. If 80% of the learners weigh at least 50 kg, how many learners weigh at least 50 kg but less than 75 kg?

6. Double percentage increase/double percentage decrease

Example

Double percentage increase

The price of a product is marked up by 15% on cost, and then by a further 10% on the first mark-up price. What is the overall percentage mark-up on cost?

Worked solution

Method

Let the original cost price = 100
After the 1st mark-up, selling price = 100 + (15% of 100) = 115
After the 2nd mark-up, selling price = 115 + (10% of 115) = 115 + 11.5 = 126.5
Mark-up % = 126.5 − 100 = 26.5%

A short-cut method

Add the two percentage increases (as decimals)   0.15 + 0.10 = 0.25
Multiply the two percentage increases (as decimals)   0.15 × 0.10 = 0.015
Finally *add* the two answers together   0.25 + 0 .015 = 0.265 (26.5%)

The final *percentage increase* will always be *more than* the sum of the two increases, since the second increase is calculated from a *larger base* (i.e. 115 instead of 100).

Example

Double percentage decrease

The price of a product is initially discounted by 20% and later by a further 10% on the discounted price. What is the overall percentage discount on the product?

Worked solution

Method

Let the original cost price = 100
After the 1st discount, the discounted selling price = [100 – (20% × 100)] = 80
After the 2nd discount, the final selling price   = [80 – (10% × 80)] = 72
Therefore the final percentage discount = 100 – 72 = 28%.

A short-cut method

Add the two percentage decreases (as decimals)   0.20 + 0.10 = 0.30
Multiply the two percentage decreases (as decimals)   0.20 × 0.10 = 0.02
Finally *subtract* the two answers from each other   0.30 – 0.02 = 0.28 (28%)

The final *percentage decrease* will always be *less than* the sum of the two increases, since the second decrease is calculated from a *smaller base* (i.e. 80 instead of 100).

1.8   Applications of percentages – word problems

1. Discounts and mark-ups

A **discount** is a *percentage decrease* calculation; a **mark-up** is a *percentage increase* calculation.

Percentage decrease calculation

Example 1
Find (a) the discount and (b) the actual amount paid when an article priced at R45 is sold at a discount of 20%.

Worked solution

Method 1	**Method 2**
Discount: 20% of R45 = R9	Let original price = 100
Actual amount paid: R45 – R9 = R36	Then discounted price = 80
	Actual discounted price $= \frac{80}{100} \times R45$
	= R36
	Discount = R45 – R36
	= R9

Percentage increase calculation

Example 2
Find (a) the mark-up and (b) the actual amount paid when an article costing R56 is sold at a mark-up of 25%.

Worked solution

Method 1	**Method 2**
Mark-up: 25% of R56 = R14	Let original price = 100
Actual amount paid: R56 + R14 = R70	Then mark-up price = 125
	Actual mark-up price $= \frac{125}{100} \times$ R56
	= R70
	Then mark-up = R70 − R56
	= R14

Exercise 14

1. (a) A radio is advertised at R160 but is sold at a discount of 15% if paid for in cash. How much does a cash customer pay for the radio?

   (b) At a sale, a TV which is priced at R800 is sold at a discount of 8%. What was the sale price of the TV?

   (c) A CD which is marked at R80 is sold at a discount of 15%. What was the sale price?

   (d) A company is offering a 12% discount on a rug which is priced at R500. What is the new selling price of the rug?

   (e) A chair was bought for R220 and sold with a mark-up of 15%. What is the selling price?

   (f) The pressure of a tyre, which was originally 180 kPa, was inflated by 11%. What is the new tyre pressure?

   (g) An account for R420 which is not paid within 30 days is charged a 20% interest. If it is settled after 30 days, how much is owing by the account holder?

   (h) A contractor forfeits 30% on any overdue contract. If a contract worth R25 000 is overdue, how much does the contractor receive?

2. (a) The price of a television set is discounted by 10%, and the reduced price is discounted by a further 10%. These successive discounts are equivalent to a single discount of what percentage?

   (b) A company reduced its workforce initially by $\frac{1}{5}$ in March due to poor trading conditions. In August it reduced its current workforce by a further $\frac{1}{4}$ when trading conditions did not improve. By what percentage has the workforce been reduced from its original size at the beginning of the year?

   (c) Increasing the original price of an article by 15% and then increasing the new price by 15% is equivalent to increasing the original price by what percentage?

   (d) The price of a dress was first discounted by a certain percentage and later by 25% of the discounted price. If the two discounts are equivalent to a single discount of 40% of the original price, what was the first discount?

3. (a) At a sale, a bed was discounted 20% and sold for R320. What was the pre-sale price?

   (b) An importer paid R520 000 for a consignment of goods. This was 30% more than the initial purchase price due to a depreciated exchange rate. What was the initial value of the consignment?

(c) After fitting a new fuel filter to a car, a 10% fuel saving is achieved. If the car now consumes 10.8 litres per 100 km, what was the car's fuel consumption per 100 km before the new fuel filter was fitted?

(d) A wholesaler adds 40% to his cost price when he sells products to a retailer. If a retailer paid R700 for a product bought from the wholesaler, what was the wholesaler's cost price?

2. Simple interest

Interest is the amount of *money earned* from an *investment* or the amount of *money charged* on money which is *borrowed*. It is always stated as a percentage per annum (% p.a.) of the money invested or borrowed.

Simple interest p.a. is always calculated on the initial amount (investment or debt) only and consequently the simple interest amount will always be the same value per annum.

To calculate simple interest, find the annual simple interest (see 1.7 part 2) and multiply this value by the number of years of the investment or the debt.

Maturity value is a general formula for calculating the maturity value (amount = A) of an initial sum invested or borrowed (principal = P) at *i*% p.a. simple interest for a term of *n* years:

$$A = P + \frac{iPn}{100} \qquad \text{or} \qquad A = P\left(1 + \frac{in}{100}\right)$$

Example 1

R600 is invested at 6% p.a. simple interest.

(a) How much interest is received after:
(i) 1 year   (ii) 2 years   (iii) 7 years?

(b) What is the maturity value of the investment after 1, 2 and 7 years, respectively?

Worked solution

Simple interest is found by calculating a percentage of a number,

i.e. for 1 year: 6% of R600 = $\frac{6}{100}$ × R600 = R36

Maturity value = R600 + R36 = R636

For 2 years: (6% p.a. of R600) × 2 = $\left(\frac{6}{100} \times 600\right)$ × 2 = R72

Maturity value = R600 + R72 = R672

For 7 years: (6% p.a. of R600) × 7 = $\left(\frac{6}{100} \times 600\right)$ × 7 = R252

Maturity value = R600 + R252 = R852

Example 2

R350 is borrowed at 8% p.a. simple interest for 5 years. How much interest is due and what is the repayment value of the loan after 5 years?

Worked solution

Apply the formula: $A = 350 + \dfrac{(8 \times 350 \times 5)}{100}$

$= 350 + 140 = R490$ (repayment value)

The simple interest due $= R490 - R350 = R140$

Exercise 15

1. Find the total simple interest earned and the maturity value (A) for:

Investment (P)	Interest rate (i% p.a.)	Term (n)
(a) R2 100	10% p.a.	3 years
(b) R6 400	15% p.a.	4 years
(c) R9 000	8% p.a.	2 years

2. Find the total simple interest charged and the repayment value (A) for:

Loan (P)	Interest rate (i% p.a.)	Term (n)
(a) Rl 200	10% p.a.	3 years
(b) R8 000	15% p.a.	5 years
(c) R4 000	9% p.a.	2 years

3. Compound interest

Compound interest is interest earned (or charged) on previously earned (or charged) interest. This means that it is *interest on interest.*

The sum of all prior interest earned (or charged) is added to the initial sum of the investment (or loan) (principal) before interest for the next period is calculated.

Thus

at the end of period 1 $A_1 = P\left(1 + \dfrac{i}{100}\right)$ $= P\left(1 + \dfrac{i}{100}\right)^1$

at the end of period 2 $A_2 = P\left(1 + \dfrac{i}{100}\right)\left(1 + \dfrac{i}{100}\right)$ $= P\left(1 + \dfrac{i}{100}\right)^2$

at the end of period 3 $A_3 = P\left(1 + \dfrac{i}{100}\right)\left(1 + \dfrac{i}{100}\right)\left(1 + \dfrac{i}{100}\right) = P\left(1 + \dfrac{i}{100}\right)^3$

Thus, the general formula for the maturity value (A) of an investment or a loan (P) for n years at i% p.a. *compound interest* is:

$$A = P\left(1 + \dfrac{i}{100}\right)^n$$

Note: A calculator is needed for compound interest calculations.

Example 1

What is the maturity value of an investment of R200 after 3 years at 10% p.a. compound interest?

Worked solution

(Maturity value) $A = 200\left(1 + \frac{10}{100}\right)^3 = 200 \ (1.1)^3 = 200(1.331) = R266.20$

When the **compounded period is less than one year**, the following adjustments to the formula are made:

1. *Divide* the *interest rate p.a.* by the number of periods within the year over which interest is compounded (i.e. by 2 if half-yearly; by 4 if quarterly; by 12 if monthly).

2. *Multiply* the *term (n)* by the same number of periods within the year over which interest is compounded.

3. Apply the revised interest rate and term to the formula.

The compound interest formula for compounding periods of less than one year is:

$$A = P\left[1 + \frac{\left(\frac{i}{k}\right)}{100}\right]^{nk}$$

Note: $k = 2$ if compounded half-yearly; $k = 4$ if compounded quarterly and $k = 12$ if compounded monthly.

Example 2

What is the maturity value of an investment of R500 after 3 years at 10% p.a. where interest is *compounded half-yearly*?

Worked solution

Divide interest rate p.a. of 10% by 2 (i.e. 5% interest earned every half-year). Also multiply the term of 3 years by 2 (i.e. 6 half-yearly periods).

Then (maturity value) $A = 500\left(1 + \frac{5}{100}\right)^6 = 500 \ (1.34) = R670.05$

Exercise 16

1. Use a calculator to find the interest earned and maturity value on the following investments:

Sum invested (P)	Interest rate p.a.	Term	Compounded
(a) R2 000	10% p.a.	2 years	annually
(b) R500	20% p.a.	3 years	annually
(c) R50	6% p.a.	3 years	half-yearly
(d) R1 000	8% p.a.	5 years	quarterly
(e) R250	12% p.a.	3 years	every 4 months
(f) R800	12% p.a.	2 years	monthly

2. Solve:

(a) Mr A bought a radio for R360 on hire purchase. He paid a deposit of 30% and the balance, on which he paid simple interest of 10% p.a., in 12 monthly instalments. How much was the deposit? How much was each monthly instalment? What was the total amount he paid for the radio?

(b) In a box of 150 glasses, 20% were broken. How many unbroken glasses were there?

(c) A college has 1 850 students. On a particular day, 4% were absent. How many students were present on the day?

(d) 3 200 people were eligible to vote in a by-election. 1 280 voted for party ABC, 800 voted for party DEF, 960 for party XYZ and the rest did not vote.

  (i) What percentage supported each party?

  (ii) What percentage did not vote?

(e) The number of people employed by a firm increased from 240 to 280. By what percentage did the workforce increase?

(f) A lady buys a car for R64 000 and sells it for R56 000 a year later. Calculate the percentage loss.

(g) A grocer buys 12 pockets of oranges at R15 each. He gives 2 pockets away to charity and sells the rest at R20 per pocket. What is his overall percentage profit margin?

(h) A worker's output was 36 units per hour on a machine. The old machine was replaced with a new one and the worker's output went up to 42 units per hour. What was the percentage increase in output?

(i) A jacket costs R180 to make and a retailer adds 25% to get the marked price. During a sale, all goods are labeled 'Sale price 10% off marked price'. What was the marked price before the sale? How much did a customer pay for the jacket during the sale? What percentage profit margin did the retailer make on the jacket which was sold during the sale?

(j) The insurance premium for a car owned by Mrs X is normally R720 per month. With a no-claim bonus, the premium is reduced by 35%. What does Mrs X pay if she has had no claims?

(k) A restaurant adds a 10% service charge on a bill and an additional 5% to the service charge for *large* tables (of 10 or more customers). What is the overall percentage increase on a bill for a customer with a *large* table?

(l) Last year a club had 45 members, each paying R50 annual subscription. This year membership increased to 60 members each paying R60 annual subscription. By what percentage has membership increased? By what percentage has total annual subscription increased?

(m) A store bought 50 articles for R400 and sold them all at R12 each. What was the total profit? What percentage profit was made on the sale of all the articles?

(n) At blast-off a rocket which weighs 8 000 kg is loaded with fuel weighing 40 000 kg. What is the mass (total weight) of the rocket when 15% of the fuel has been used?

(o) A house which was bought for R270 000 is sold for R350 000 a few years later. If the estate agent's commission was 7% of the selling price and there were a further R1 200 in selling expenses, what percentage capital gain did the seller make on the sale?

1.9   Ratios

Ratios are a disguised form of fractions. The ratios identify the fraction (or part) of a whole.

1. Finding the value of each part

Example

R80 is to be divided in the ratio of 2 : 3 between two persons. How much does each person receive?

Worked solution

Method

1. Add the parts of the ratio to make up the whole: (2 + 3 = 5 parts = whole)
2. Express each part as a fraction of the whole: $\left(\frac{2}{5} \text{ and } \frac{3}{5}\right)$
3. Find each fraction of the value given: $\left(\frac{2}{5} \text{ of } 80 \text{ and } \frac{3}{5} \text{ of } 80\right)$

    The first person receives $\frac{2}{5} \times$ R80   = R32

    The second person receives $\frac{3}{5} \times$ R80 = R48

2. Finding the whole when given a part

When given a value for one of the parts, find the values for the other parts or the whole.

Example

A box contains nails and screws in the ratio of 5 : 3. If there are 20 nails in the box, how many screws does the box contain?

Worked solution

Method 1   Scaling the ratios

                    Nails : Screws
Ratio   5 : 3
Numbers   20 : ?
Multiply both sides of the ratio by 4 $\left(= \frac{20}{5}\right)$ to equate the ratio of nails to 20.
Ratio × 4   20 : 12
Hence there are 12 screws in the box.

Method 2   Using proportions

Total of ratios = 5 + 3 = 8. Hence nails = $\frac{5}{8}$ and screws = $\frac{3}{8}$.

Now the fraction of $\frac{5}{8}$ = 20 nails

Then the fraction of $\frac{3}{8}$ = x screws, where $x = \frac{\frac{3}{8}}{\frac{5}{8}} \times 20 = 12$ screws

Exercise 17

1. Solve:

   (a) Share R120 in the ratio of 3 : 5.

   (b) Divide R6 300 in the ratio of 2 : 7.

   (c) Share 520 sweets in the ratio of 2 : 3 : 5.

   (d) Divide R660 in the ratio of 3 : 2 : 5 : 5.

   (e) A 520 g cake is divided in the ratio of 4 : 2 : 7. How much does the smallest portion weigh?

   (f) R1 440 is divided between the eldest child and her younger twin brothers in the ratio of 6 : 5 : 5. How much does each twin receive?

   (g) The amounts of time that three programmers, A, B and C, worked on an assignment are in the ratio of 1 to 2 to 5, respectively. If they worked a combined total of 112 hours, how many more hours did programmer C work on the assignment compared to the programmer A?

   (h) A fruit-salad mixture consists of apples, peaches and grapes in the ratio of 6 : 5 : 2, respectively, by weight. If 52 kg of the mixture is prepared, the mixture includes how many more kg of apples than grapes?

   Hint: When a ratio question uses words like 'A has a *third less* than B ...', or 'X receives a *quarter more* than Y ...', or 'P has *20% less* than Q ...', then give a value of 1 (or 100%) to the object that does not depend on any other objects (i.e. look for the independent event). Then work out the ratio of the other objects using the percentage or fraction change from 1 (or 100%).

Example 1

'A has a *third less* than B'.

Then let $\quad$ B = 1, and A = $\left(1 - \frac{1}{3}\right) = \frac{2}{3}$

Then ratio: $\quad$ A : B is $\frac{2}{3}$: 1 $\quad$ or $\quad$ 2 : 3

Example 2

'X receives a *quarter more* than Y'.

Then let $\quad$ Y = 1, and X = $1 + \frac{1}{4} = 1\frac{1}{4}\left(\text{or } \frac{5}{4}\right)$

Then ratio: $\quad$ X : Y is $\frac{5}{4}$: 1 $\quad$ or $\quad$ 5 : 4

Example 3

'P has 20% *less* than Q'.

Then let $\quad$ Q = 100%

Then $\quad$ P = (100 – 20)% = 80%

Then ratio: $\quad$ P : Q is 80 : 100 $\quad$ or $\quad$ 4 : 5

2. Solve:

   (a) Tim and Joe share R160 in the ratio of 2 : 3. How much less does Tim get than Joe?

   (b) Divide R96 between A and B so that B has twice as much as A.

   (c) Share R168 so that A has a third more than B.

   (d) Share R460 such that A has half as much as C, who has a quarter more than B.

   (e) Share R54 so that B has three times as much as C, who has half as much as A.

3. Solve:

   (a) In a class, the ratio of boys to girls is 3 : 2. If there are 12 boys in the class, how many girls are there?

   (b) A box has a ratio of pens to pencils of 5 : 3. How many pencils are there if there are 15 pens in the box?

   (c) A sum of money is divided in the ratio of 4 : 5. If the smaller amount is R160, what is the larger amount?

   (d) An alloy consists of copper, zinc and tin in the ratio of 1: 3 : 4. If there is l0 g of copper in the alloy, find the weights of zinc and tin.

   (e) On a farm, the ratio of cows to horses to sheep is 7 : 2 : 3. If there are 48 sheep, how many more cows are there than horses?

   (f) The ratio of women to men in a bowling club is 9 : 5. If there are 45 women in the club, how many members does the club have altogether?

   (g) Three business partners, P, Q and R, agree to divide their total profit for a certain year in the ratio of 2 : 5 : 8, respectively. If P's share was R4 000, how much more did R receive than Q?

   (h) The ratio of women to children in a group is 5 : 2. The ratio of children to men in this group is 5 : 11. If there are 50 women in the group, how many men are there?

   (i) The ratio of women to children in a group is 5 : 2. The ratio of children to men in this group is 5 : 11. If there are *less than* 30 women in the group, how many men are there?

   (j) In a company, the ratio of management to workers is 1 : 9. If $\frac{2}{3}$ of the workers are female, and $\frac{1}{4}$ of the management are female, then what fraction of the combined management and worker complement are female?

   (k) In an election, candidate X received $\frac{1}{3}$ more votes than candidate Y. Candidate Y received $\frac{1}{4}$ fewer votes than candidate Z. If candidate Z received 2 400 votes, how many votes did candidate X receive?

1.10 Proportion

A **proportion** is a way of comparing two fractions to find a missing value in one of the fractions that will make them equivalent.

Proportions are either *direct* (when the ratio of the two fractions is the same) or *indirect* (when the ratio of the two fractions is inverse).

a. Direct proportions

Two quantities are in direct proportions when an increase in one is in the same ratio as an increase in the other. For example, if 5 apples cost R15, then 10 apples cost R30. Since each apple costs R3, the ratio of apples to cost is 1 : 3.

Example

If 4 m of cloth costs R20, what will 10 m of cloth cost?

Worked solution

Method 1

Make a proportions statement to form *equivalent fractions*:
If   4 m of cloth cost R20
then 10 m of cloth cost Rx i.e. $\frac{10}{4} = \frac{x}{20}$ as equivalent fractions
Thus $x = \left(\frac{10}{4} \times \text{R20}\right) = \text{R50}$

Method 2

Find the cost of 1 unit (i.e. 1 m of cloth) and multiply the unit cost by the new quantity desired (i.e. 10 m of cloth).
If 4 m of cloth costs R20, then 1 m of cloth will cost R5 $\left(\text{i.e. R}\frac{20}{4}\right)$
Therefore 10 m of cloth will cost $10 \times \text{R5} = \text{R50}$

b. Indirect proportions

Two quantities are in *indirect* proportion to each other when an increase in one quantity leads to a decrease in the other quantity by a fixed ratio (i.e. the ratios between the two quantities is inverted).

For example, if 3 workers complete a task in 12 hours, then 6 workers (working at the same rate) will complete the same task in half the time, i.e. 6 hours ($x = \frac{3}{6} \times 12$). More workers will take less time to do the same job.

Hint: First work out the total value to complete a task (e.g. number of men working x days to complete task = total work days for the complete task). Now to achieve this same total value when one of the input values (e.g. fewer men) has changed, divide this total value by the changed input value.

Example

A car, travelling at a constant speed of 60 km/h, takes 4 hours to complete a journey. How long will it take to cover the same distance at 80 km/h?

Worked solution

The journey will take *less time* at the *faster speed* of 80 km/h.
Hence this is an indirect proportion sum.
First find the total value (i.e. distance travelled) of the journey at 60 km/h for 4 hours = 240 km.
Now, to find how long it would take to complete the 240 km journey at 80 km/h, divide 240 by 80. Thus $\frac{240}{80} = 3$ means that it will take 3 hours to complete the same journey.

Exercise 18

1. Solve the following questions related to direct proportion:

    (a) If 7 books cost R84, what is the total cost of 11 books? (All books cost the same.)

    (b) 20 workers produce 500 articles in a week. How many articles would 4 workers produce in a week?

    (c) If a runner takes 56 minutes to run 8 km, how long will he take to finish a 20 km race?

    (d) A set of 6 ceramic tiles cost R42. What will you pay for 54 ceramic tiles?

    (e) A machine fills 2 000 bottles in 10 minutes. How many bottles will it fill in 7 minutes?

    (f) A plane flies 50 km in 15 minutes. How long will it take to fly 350 km?

    (g) If a $2\frac{1}{2}$ m long plastic pipe costs R4.50, how much will a 4 m length of plastic pipe cost?

    (h) An aircraft uses 150 litres of fuel to fly 375 km. How much fuel is required for a journey of 500 km?

    (i) 7 bicycles cost R6 230. What is the cost of 3 bicycles? How many bicycles can be bought for R9 790?

    (j) If carpeting costs R3 500 for 20 m², how much would it cost to carpet a room 4 m by 6 m?

    (k) 6 identical machines running at the same constant rate can produce a total of 270 bottles per minute. At this same rate, how many bottles could 10 such machines produce in 4 minutes?

2. Solve the following questions related to indirect proportion:

    (a) If 4 men complete a job in 10 hours, how long will it take 5 men to complete the same job?

    (b) A pump can fill a tank in 21 hours at a constant rate (in litres per hour). How long will it take the pump to fill the tank if its rate of flow increased by 50%?

    (c) 24 horses eat a lorry load of hay in 10 days. If a third of the horses were sold, how many days would a lorry load of hay feed the remaining horses?

1.11 Speed, distance, time

Note 1 This section (Speed, distance, time) and the next section (Rates of work/ flow) are identical in the way problems are solved.

Note 2 To compare or combine different speeds (or rates of flow) always convert each speed (or rate of flow) to the *same unit of time* (e.g. km per *hour*; litres per *minute*).

As an aid to memory, remember the triangle:

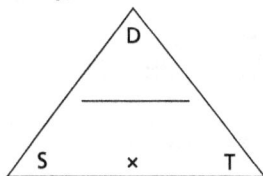

D = distance
S = speed
T = time

1. To find *distance,* cover D so that D = S × T   (i.e. distance = speed × time)

2. To find *speed,* cover S so that S = $\frac{D}{T}$ $\left(\text{i.e. speed} = \frac{\text{distance}}{\text{time}}\right)$

3. To find *time,* cover T so that T = $\frac{D}{S}$ $\left(\text{i.e. time} = \frac{\text{distance}}{\text{speed}}\right)$

Values are always given for two of the three terms. You then need to find the value of the missing term.

Example 1: Finding speed	**Example 2: Finding distance**
A car travels 450 km in 5 hours. What was the average speed of the car?	A car takes 8 hours travelling at an average speed of 90 km/h to complete a journey. What was the length of the journey?
Worked solution	**Worked solution**
Speed = $\frac{D}{T}$ = $\frac{450}{5}$ = 90 km/h	Distance = S × T = 90 × 8 = 720 km

Example 3: Finding Time

A car completes a journey of 1 188 km travelling at an average speed of 99 km/h. How long did the journey take?

Worked solution

Time = $\frac{D}{S}$ = $\frac{1\,188}{99}$ = 12 hours

Exercise 19

1. Solve:

    (a) A car travelling at a steady speed takes 6 hours to travel 492 km. What was the speed of the car?

    (b) An aircraft flies at an average speed of 800 km/h for $3\frac{1}{2}$ hours. How far does the aircraft fly?

    (c) If a newborn baby has grown by 4 cm in 28 days, by how much does she grow on average per day?

    (d) An athlete runs at a steady speed of 5 metres per second (m/s) for 100 seconds. How far does she run?

    (e) A rocket can fly at 1 000 km/h. If a target is 600 km away, how long will it take to reach its target?

(f) A rowing boat finished a race of 8 km at a speed of 20 km/h. How long did it take to complete the race?

(g) A car travels from A to B at an average speed of 60 km/h and returns immediately along the same route at an average speed of 80 km/h. What was the average speed (to one decimal point), for the round-trip?

(h) One athlete completes a 30 km race in 2 hours while another athlete takes $2\frac{1}{4}$ hours. What is the difference in speed (in km/h) between the two athletes?

1.12 Rates of work/flow

A rate describes units of output per time interval:

$$\text{Rate} = \frac{\text{total output units}}{\text{total time (to complete a task)}}$$

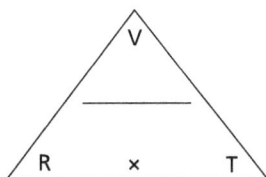

V = volume
R = rate
T = time

To find the total time to complete a task, rearrange the formula:

$$\text{Total time} = \frac{\text{total output units}}{\text{rate}}$$

Note: To find the *rate of completion* of any task, always let the complete task (V) = 1. For example, if it takes 5 hours (T = 5) to complete the task (V = 1), then, the rate (R) of task completion per hour is $\frac{1}{5}$. This means that $\frac{1}{5}$ of the task is completed every hour $\left(\text{i.e. rate} = \frac{V}{T}\right)$.

Rule 1 To find the **rate of completion (R)** for an unspecified task, always *invert* the task completion time (T) $\left(\text{i.e. } R = \frac{1}{T}\right)$:

e.g. if T = 8 hours, then *completion rate* = $\frac{1}{8}$ of task every hour.

Rule 2 If the rate per unit of time is known, then the **completion time (T)** for a task is found by inverting the rate (R) $\left(T = \frac{1}{R}\right)$:

e.g. if R = $\frac{2}{5}$ per hour, then task completion time = $\frac{1}{\left(\frac{2}{5}\right)} = \frac{5}{2}\left(2\frac{1}{2} \text{ hours}\right)$

Example 1
A factory produces 50 units in 10 hours. What is the unit rate per hour?

Worked solution

Unit rate/hour R = $\frac{V}{T} = \frac{50}{10}$ = 5 units per hour

Example 2

It takes a worker 8 hours to complete a task. What is his work rate per hour?

Worked solution

Let $V = 1$, then $R = \frac{V}{T} = \frac{1}{8}$ per hour

(i.e. he completes $\frac{1}{8}$ of the task in each hour)

Example 3

How long will it take a worker producing parts at the rate of 15 per minute, to fill an order requiring 600 parts?

Worked solution

Total time: $T = \frac{V}{R} = \frac{600}{15} = 40$ minutes

Example 4

An inlet pipe fills an empty tank in 4 hours. A second inlet pipe takes 3 hours to fill the same tank.
If both pipes are used together, how long will it take:
(a) to fill the tank completely?
(b) to fill the tank to $\frac{3}{4}$ of its capacity?

Worked solution

Rate of flow per hour from inlet pipe 1 $= \frac{1}{4}$ per hour
(i.e. every hour it fills $\frac{1}{4}$ of the tank)
Rate of flow per hour from inlet pipe 2 $= \frac{1}{3}$ per hour
(i.e. every hour it fills $\frac{1}{3}$ of the tank)
Combined rate of flow per hour from both inlet pipes $= \left(\frac{1}{4} + \frac{1}{3}\right) = \frac{7}{12}$ per hour

(a) To fill tank $\left(\text{i.e. } \frac{12}{12}\right)$, use proportions:

   $\frac{7}{12}$ of the tank is filled in 1 hour by both pipes together

   $\frac{12}{12}$ of the tank is filled in $\left(\frac{\frac{12}{12}}{\frac{7}{12}}\right) \times 1 = \frac{12}{7}$ hours $= 1\frac{5}{7}$ hours

(b) To fill $\frac{3}{4}$ of tank: $\left(\text{i.e. } \frac{9}{12}\right)$, use proportions:

   $\frac{7}{12}$ of the tank is filled in 1 hour by both pipes together

   $\frac{9}{12}$ of the tank is filled in $\left(\frac{\frac{9}{12}}{\frac{7}{12}}\right) \times 1 = \frac{9}{7}$ hours $= 1\frac{2}{7}$ hours

Exercise 20

1. Solve:

    (a) Coins are dropped into a toll box at the rate of approximately 2 cubic feet per hour. If the rectangular box is 4 feet long, 4 feet wide and 3 feet high, how long will it take to fill the box?

(b) Machine A produces bolts at a uniform rate of 120 every 40 seconds; machine B produces bolts at a uniform rate of 100 every 20 seconds. If the two machines run simultaneously, how many seconds will it take for them to produce a total of 200 bolts?

(c) If car A travels 120 km in 3 hours and car B travels 200 km in 4 hours, what is the total distance travelled by both cars in 1 hour?

(d) Worker X completes $\frac{1}{3}$ of a job in 3 hours, while worker Y completes $\frac{3}{5}$ of the same job in 4 hours.

   (i) If both workers work together on the job, what fraction of the job is completed in 1 hour?

   (ii) What percentage of the job is completed in 2 hours?

   (iii) How long would it take both workers working together to complete the whole job?

(e) A photocopier makes 2 copies every $\frac{1}{3}$ second. At the same rate, how many copies does it make: (i) in 4 minutes (ii) in 2 hours?

(f) Three machines individually can do a certain job in 4, 5 and 6 hours, respectively. What is the greatest part of the job that can be done in one hour by two of the machines working together at their respective rates?

(g) An empty pool being filled with water at a constant rate takes 8 hours to fill to $\frac{3}{5}$ of its capacity. How much more time will it take to finish filling the pool?

(h) One inlet pipe fills an empty tank in 5 hours. A second inlet pipe fills the same tank in 3 hours. If both pipes are used together, how long will it take to fill $\frac{2}{3}$ of the tank?

(i) Sam and Bill take $\frac{5}{8}$ of the time together to complete a task that Sam would do in 6 hours on his own. How long will it take Bill to do the same task on his own?

(j) Two pumps simultaneously fill a pool in 4 hours when pumping at constant rates. Pump 1 is 1.5 times faster. How long does pump 1 take to fill the pool by itself?

1.13 Quick-and-dirty arithmetic

Mental arithmetic can be performed faster by using a few short-cut arithmetic tricks – called *quick-and-dirties*. The idea is to look for easier ways to perform arithmetic operations. Here are a few tips:

1. Splitting terms

Split the terms of the expression into more easily manageable parts, usually involving multiples of 10 or 5.

Example 1	Example 2
Calculate 28 × 15	Calculate 19 × 14

Worked solution	**Worked solution**
28×15	19×14
$= (28 \times 10) + \frac{1}{2}(28 \times 10)$	$= (20 \times 14) - (1 \times 14)$
$= 280 + 140 = 420$	$= 280 - 14 = 266$

Use your imagination to decide what breakdown of the expression will be easier and faster to work with.

2. Finding percentages of numbers

There are **two** *quick-and-dirty* methods that can be considered here. The choice depends on whether the numbers lend themselves to easy manipulation.

Method 1
Split the percentage into easier parts, which can be worked out separately.

Example 1
Calculate 11% of 260.

Worked solution

Calculate 10% and 1%, respectively, of 260 and add the results:
i.e. 10% of 260 = 26; 1% of 260 = 2.6
Now add 26 + 2.6 = 28.6

Example 2
Calculate 75% of 240.

Worked solution
75% is $\frac{3}{4}$ as a fraction. Find $\frac{1}{4}$ of 240 and multiply by 3:
i.e. $\frac{1}{4}$ of 240 = 60, then $3 \times 60 = 180$.

Example 3
Calculate 125% of 64.

Worked solution
Split 125% into 100% + 25%. Work out each percentage separately and add together, i.e. 100% of 64 = 64; 25% of 64 = $\frac{1}{4}$ of 64 = 16. Then find 64 + 16 = 80.

Method 2
Find an (easy) factor that converts 100 into the number. Then multiply (or divide) the percentage by this factor.

Example 1	**Example 2**
Calculate 18% of 50.	Calculate 46% of 200.

Worked solution

$\frac{100}{2}$ = 50, therefore divide 18% by 2 to give the answer of 9.

Worked solution

100 × 2 = 200, therefore multiply 46% by 2 to give the answer of 92.

Example 3

Calculate 56% of 25.

Example 4

Calculate 70% of 800.

Worked solution

$\frac{100}{4}$ = 25, therefore divide 56% by 4 to give the answer of 14.

Worked solution

100 × 8 = 800, therefore multiply 70% by 8 to give the answer of 560.

3. Testing whether a number is divisible by 2, 3, 4, 5, 6, 8, 9 and 11

a. Divisible by 2

The number must be even (i.e. it must end in either 0, 2, 4, 6 or 8).

b. Divisible by 3

The sum of the digits must be divisible by 3.

Example 1

327 is divisible by 3, since 3 + 2 + 7 = 12 is divisible by 3.

Example 2

593 is not divisible by 3, since 5 + 9 + 3 = 17 is not divisible by 3.

c. Divisible by 4

The last two digits of the number must be divisible by 4.

Example 1

3 716 is divisible by 4, since 16 is divisible by 4.

Example 2

826 is not divisible by 4, since 26 is not divisible by 4.

d. Divisible by 5

The number must end in either a 0 or 5 (e.g. 15, 45, 60, 105, 970, etc.).

e. Divisible by 6

The number must be *even* and the sum of all the digits must be divisible by 3.

Example 1

3 078 is divisible by 6, since it is even and 3 + 0 + 7 + 8 = 18 is divisible by 3.

Example 2

826 is not divisible by 6. While it is even, 8 + 2 + 6 = 16 is not divisible by 3.

f. Divisible by 8

The last three digits of the number must be divisible by 8.

Example 1

7 664 is divisible by 8, since 664 is divisible by 8.

Example 2

56 172 is not divisible by 8, since 172 is not divisible by 8.

g. Divisible by 9

The sum of all the digits must be divisible by 9.

Example 1

11 043 is divisible by 9, since 1 + 1 + 0 + 4 + 3 = 9 is divisible by 9.

Example 2

8 618 is not divisible by 9, since 8 + 6 + 1 + 8 = 23 is not divisible by 9.

h. Divisible by 11

The sum of all pairs of digits (starting from the units side of the number) must be divisible by 11.

Example 1

759 is divisible by 11, since 07 + 59 = 66 is divisible by 11.

Example 2

5 612 is not divisible by 11, since 56 + 12 = 68 is not divisible by 11.

4. Approximating answers (rounding off to the nearest convenient number)

Rounding

When rounding off a decimal number to k decimal places:

1. Truncate at the kth decimal place if the $(k + 1)^{th}$ digit before rounding is ≤ 4.
2. Round up at the kth decimal place if the $(k + 1)^{th}$ digit before rounding is ≥ 5.

Example 1

Round 34.473 off to the first decimal place.

Worked solution

34.473 = 34.5
(the 2nd decimal digit is ≥ 5)

Example 2

Round 0.2382 off to the first decimal place.

Worked solution

0.2382 = 0.2
(the 2nd decimal digit is ≤ 4)

Example 3

Round 864.1482 off to the first decimal place.

Worked solution

864.1482 = 864.2

(Since the 3rd decimal digit = 8, the 2nd decimal digit rounds up to 5, resulting in the 1st decimal digits being rounded up to 2.)

Approximating

To approximate calculations, round off numbers to the *closest integers* or *easy-to-work-with* numbers.

Example 1

Approximate the answer: $\frac{7.79}{1.96}$.

Worked solution

Round to $\frac{7.8}{2}$ = 3.9
(actual value = 3.97)

Example 2

Approximate the answer: 816 × 0.104.

Worked solution

816 × 0.104 (round to 1 d.p.)
816 × 0.1 = 81.6
(actual value = 84.86)

5. Calculating decimals for fractions of $\frac{1}{9}$ and $\frac{1}{11}$

a. $\frac{1}{9}$ fractions

Repeat the numerator value after the decimal point (it is a recurring decimal):

$\frac{1}{9}$ = 0.1111111 (recurring)

$\frac{2}{9}$ = 0.2222222 (recurring)

$\frac{3}{9}$ = 0.3333333 (recurring)

$\frac{4}{9}$ = 0.4444444 (recurring)

$\frac{5}{9}$ = 0.5555555 (recurring)

$\frac{6}{9}$ = 0.6666666 (recurring)

$\frac{7}{9}$ = 0.7777777 (recurring)

$\frac{8}{9}$ = 0.8888888 (recurring)

b. $\frac{1}{11}$ fractions

Multiply the numerator by 9. Then repeat the answer as a two-digit number after the decimal point (it is also a recurring decimal):

$\frac{1}{11}$ = 0.090909 (recurring) $\frac{6}{11}$ = 0.545454 (recurring)

$\frac{2}{11}$ = 0.181818 (recurring) $\frac{7}{11}$ = 0.636363 (recurring)

$\frac{3}{11}$ = 0.272727 (recurring) $\frac{8}{11}$ = 0.727272 (recurring)

$\frac{4}{11}$ = 0.363636 (recurring) $\frac{9}{11}$ = 0.818181 (recurring)

$\frac{5}{11}$ = 0.454545 (recurring) $\frac{10}{11}$ = 0.909090 (recurring)

An interesting observation: the decimal fractions from $\frac{6}{11}$ to $\frac{10}{11}$ are the transpose (reverse) of the decimal fractions from $\frac{1}{11}$ to $\frac{5}{11}$.

6. Finding the square of any number

To find the square of a number:

1. Find a number that, when added and subtracted from the number to be squared, results in two easy-to-multiply numbers.
2. Find the product of these two new numbers and add the square of the added (and subtracted) number.

This may seem confusing, but let's look at it in action.

Example 1

Find the square of 16 (i.e. 16^2).

Worked solution

(We know the answer is 256!)

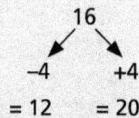

$$16$$
$$-4 \quad\quad +4$$
$$= 12 \quad = 20$$

Choose 4 as the number to add and subtract from the number to be squared. This gives $12 \times 20 = 240$ (i.e. find the product of these two new numbers). Finally $240 + 4^2 = 256$ (i.e. add the square of the added (subtracted) number to the product).

Example 2

Find the square of 33 (i.e. 33^2).

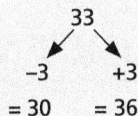

Exercise 21

1. Use a short-cut method to find:

   (a) 11% of 460   (b) 9% of 26   (c) 21% of 150

   (d) 75% of 280   (e) 125% of 120   (f) 49% of 810

2. Which of the following numbers are divisible by 3? If not, what would be the closest number that would be divisible by 3?

   (a) 34   (b) 54   (c) 72   (d) 68

   (e) 124   (f) 338   (g) 504   (h) 1 234

3. Which of the following numbers are divisible by 6? If not, what would be the closest number that would be divisible by 6?

   (a) 138   (b) 77   (c) 272   (d) 671

   (e) 124   (f) 309   (g) 504   (h) 1 932

4. Which of the following numbers are divisible by 9? If not, what would be the closest number that would be divisible by 9?

   (a) 108   (b) 774   (c) 233   (d) 668

   (e) 414   (f) 918   (g) 1 404   (h) 1 934

5. Which of the following numbers are divisible by 11? If not, what would be the closest number that would be divisible by 11?

   (a) 77   (b) 91   (c) 187   (d) 351

   (e) 517   (f) 919   (g) 2 431   (h) 6 809

6. Approximate the answers to the following:

   (a) $\frac{27.03}{5.1}$ (b) $\frac{59.78}{9.8}$ (c) $\frac{13.12}{1.92}$ (d) $\frac{18.05}{0.97}$

   (e) $\frac{705}{7.102}$ (f) $\frac{1.078}{0.098}$ (g) $\frac{20.88}{2.88}$ (h) $\frac{2363.8}{99}$

   (i) 32 × 98   (j) 2.1 × 5.01   (k) 25.11 × 19.76

   (l) 974 × 0.11   (m) 19.22 − 14.28

   (n) 207.12 + 18.53   (o) 664 × 0.51

7. Find the square of the following numbers (use a short-cut method):

   (a) 18   (b) 22   (c) 35

   (d) 63   (e) 55   (f) 97

CHAPTER

2

FUNDAMENTAL ALGEBRA

Important note: The rules of algebra are exactly the same as the rules of arithmetic. The only difference is that symbols (letters) replace numbers.

2.1 Basic concepts: terms and expressions

A **term** consists of numbers and symbols joined by multiplication and division signs:

e.g. $\quad x \quad\quad 3xy \quad\quad 14x^3y^2 \quad \dfrac{3}{(x+y)} \quad (x+2)(x-3) \quad \dfrac{12xy^2}{(3y-4)}$

Note: The term xy is the same term as yx, but x^2y; xy^2 and $(xy)^2$ are not the same terms.

An algebraic **expression** consists of terms joined by addition and subtraction signs only:

e.g. $\quad 4x + 3x \quad\quad 3a + 4b - 5c \quad\quad 4x^2 + 2x - 4 \quad\quad -4p + 2q - 7r + 2$

Common errors in adding and multiplying like terms

(a) When like terms, such as y, are *added* k times, i.e. $y + y + y + \ldots + y$, the answer is ky:

e.g. $y + y + y + y + y = 5y$ (in numbers: $7 + 7 + 7 + 7 + 7 = 5(7) = 35$)

(b) When like terms, such as y, are *multiplied* together k times, i.e. $y \cdot y \cdot y \cdot y \ldots y$, the answer is y^k (a common error is to say $k \times y$):

e.g. $y \cdot y \cdot y \cdot y \cdot y = y^5$ (in numbers: $7 \cdot 7 \cdot 7 \cdot 7 \cdot 7 = 7^5 = 16\,807\ [\neq 5(7) = 35]$)

2.2 Forming algebraic expressions

When a word problem uses symbols to represent quantities (or numbers), the solution is formulated as an algebraic expression.

The rules of arithmetic are used to construct the algebraic form of the word problem solution.

To help you form algebraic expressions, use substitution of numbers for symbols to understand how to solve the word problem. Once the logic has been worked out, replace the numbers with the symbols.

Example 1

A car travels a distance of x km in p hours. What was the car's average speed per hour for the journey?

Worked solution

To understand the problem, let $x = 100$ km and $p = 2$ hours.

Using arithmetic: speed $= \dfrac{100}{2} = 50$ km/h; thus algebraically: speed $= \dfrac{x}{p}$ km/h

Example 2

A customer bought p units of product A at q cents per unit, and r units of product B at t cents per unit. What was the average price of each unit bought?

Worked solution

To understand the problem, let $p = 10$ units of A; $q = 20c$ per unit of A; $r = 6$ units of B; and $t = 30c$ per unit of B.

Using arithmetic: average price $= \dfrac{[(10 \times 20c) + (6 \times 30c)]}{(10 + 6)}$

Thus algebraically: average price per unit $= \dfrac{(pq + rt)}{(p + r)}$

Exercise 22

Find algebraic expressions or terms for each of the following:

(a) If Susan was 24 years old x years ago, how old will she be in z years' time?

(b) A group is composed of w women and m men. If 3 women and 2 men join the group, what is the probability of randomly selecting a woman from this enlarged group?

(c) A retailer bought 100 identical shirts at a total cost of s rand. If each shirt was sold at 50% above the unit cost per shirt, what was the selling price for each shirt?

(d) How many minutes does it take Paul to type y words if he types at a rate of x words per minute?

(e) Sipho has x rand more than Mpumi. Together they have y rand. How much does Mpumi have?

(f) If Mandla takes 11 seconds to run y metres, how many seconds will it take him to run x metres at the same speed?

(g) The sum of the ages of Doris and Fred is y years. If Doris is 12 years older than Fred, how old will Fred be y years from now?

(h) The cost to rent a bus for a trip is x rand. This amount is to be shared equally among the people taking the trip. If 10 people take the trip rather than 16 people, how much more will it cost each person?

(i) Alf has x rand more than Beth has, and together they have a total of y rand. How many rand does Beth have?

(j) A dealer bought 100 identical batteries at a total cost of q rand. If each battery was sold at 50% above its cost price, what was the selling price of each battery (in q terms)?

2.3   Operations on algebraic expressions

1. Grouping like terms

Rule   Group *like terms* together and write them as a single term:
e.g. $7x - 4x + 2x = 5x$.

Unlike terms (e.g. 5*x*, 8*y*, 4*p*, etc.) cannot be grouped together:
e.g. $6x + 2y + x - 4x - 7y = (6x + x - 4x) + (2y - 7y) = 3x - 5y$.

Example 1	**Example 2**
Simplify $5x + 4y - 2x - 6y$.	Simplify $9a - 14b + c - 2a + 8b$.
Worked solution	**Worked solution**
$5x + 4y - 2x - 6y$	$9a - 14b + c - 2a + 8b$
$= (5x - 2x) + (4y - 6y)$	$= (9a - 2a) + (-14b + 8b) + c$
$= 3x - 2y$	$= 7a - 6b + c$

2. Removing brackets

When removing brackets follow these steps:

1. Multiply the term outside a bracket with every term inside the bracket.

2. If a negative sign is in front of a bracket, change the sign of each term inside the bracket.

3. Simplify the expression by grouping like terms.

Example 1	**Example 2**
Simplify $2x + 3(4x - 2)$.	Simplify $9a - 2(5 - 4a)$.
Worked solution	**Worked solution**
$2x + 3(4x - 2)$	$9a - 2(5 - 4a)$
$= 2x + 12x - 6$	$= 9a - 10 + 8a$
$= 14x - 6$	$= 17a - 10$

3. Multiplying a bracket by a bracket

When multiplying a bracket by a bracket, follow these steps:

1. Multiply each term from the first bracket with each and every term from the second bracket.
   Take care when *multiplying with minus (–) sign numbers.*

2. Simplify the expression by grouping like terms.

Example 1	**Example 2**
Simplify $(3x - 2)(x + 3)$.	Simplify $(a - 4)(a + 4)$.
Worked solution	**Worked solution**
$(3x - 2)(x + 3) = 3x(x + 3) - 2(x + 3)$	$(a - 4)(a + 4) = a(a + 4) - 4(a + 4)$
$= 3x^2 + 9x - 2x - 6$	$= a^2 + 4a - 4a - 16$
$= 3x^2 + 7x - 6$	$= a^2 - 16$
See **Note 1** on the next page.	See **Note 2** on the next page.

Note 1   Note the similarity in terminology between arithmetic and algebra. In arithmetic, since 4 × 6 = 24, we say that 4 and 6 are the *factors* of 24. In algebra, since $(3x - 2)(x + 3) = 3x^2 + 7x - 6$, we also say that the brackets $(3x - 2)$ and $(x + 3)$ are the *factors* of the expression $(3x^2 + 7x - 6)$, which represents a number.

Note 2   If an expression has only *two terms*, both of which are *perfect squares* (e.g. a^2 and 16), and the terms are separated by a *minus sign,* then the expression is called a **difference between two squares**.

The *two factors* of a *difference between two squares* expression are always the *square root of each term*, separated in *bracket one* by a *plus sign* and separated in *bracket two* by a *minus sign*.

4. Substitution

With *substitution, numbers replace* the *symbols* in an expression. The expression, when simplified, then equals a specific number.

Example 1	Example 2
Find the value of the expression $4x - 1$ when $x = 2$.	Find the value of the expression $5x^2 + 4x$ when $x = 2$.
Worked solution	**Worked solution**
$4x - 1$ $= 4(2) - 1$ (Replace x with 2) $= 7$	$5x^2 + 4x$ $= 5(2)^2 + 4(2)$ (Replace x with 2) $= 20 + 8$ $= 28$

Exercise 23

1.   Find the simplest form of each expression:

(a) $5x - 6y + 3x - 2y$ (b) $7q + 12r + 3q - 15r$

(c) $11a - 2b + 2 - 4a - 9b$ (d) $12d - 9d^2 + 2d + 9d^2 - 9d$

(e) $2a^2 + ab - 5ba + 7b^2$ (f) $-p^2 + 4pq - 2q^2 + 5p^2 - 5pq$

(g) $8x^2y + 2xy^2 + 2x^2 - 7yx^2 + 4x^2$ (h) $-3abc + 6ab - 4cba - 6bac$

2.   Find the simplest form of each expression by removing the brackets first:

(a) $3m + 4(m + 3)$ (b) $2t + 5(4t - 5)$

(c) $6r - 3(3r - 5)$ (d) $v - 3(5v - 2) - 4$

(e) $-4x - 5(4x - 3) + 2$ (f) $8 - 3(5m - 2) - 4m$

(g) $9 - 2(3y + 4) - 3y$ (h) $5x + 2y - 4(x - 1) + x$

(i) $8a - 2a(3 - a) + a^2$ (j) $y^2 + 2y(3y - 4) - 6$

(k) $5x - 2(x + 4y - 2)$ (l) $2a + 3(2a - 4)$

(m) $-2(2x - 3y) - 4x + y$

3. Find the product of the following factors and simplify the resulting expression. Indicate which expressions are a *difference between two squares*.

(a) $(3m + 3)(4m + 3)$ (b) $(2t + 5)(4t - 5)$ (c) $(4r - 3)(3r - 5)$

(d) $(5c + d)(5c - d)$ (e) $(4 - 3y)(2y + 1)$ (f) $(x - 2y)(4x + 3y)$

(g) $(8p - q)(9p - 2q)$ (h) $(2a - 4b)(a - b)$ (i) $(b - 2)(b + 2)$

(j) $(7m - 4)(7m + 4)$ (k) $(v + 3)(5v + 4)$ (l) $(4p - q)(p - 3q)$

(m) $(y + 3)^2$ (n) $(2p - q)^2$ (o) $(x + 3)(x - 2)$

(p) $(b + 2)(b - 4)$ (q) $(2k - 3)(2k + 3)$ (r) $(3d + 2)(2d - 1)$

(s) $(x + 2)^2$ (t) $(4a - 5b)^2$

4. Find the value of each expression for the given number(s) of the unknown:

(a) $9x - 5$ for $x = 3$ (b) $2x^2 + 4$ for $x = 2$

(c) $14y - 3$ for $y = \frac{1}{2}$ (d) $3v^2 + 2v$ for $v = 4$

(e) $\frac{8x}{3} + 2$ for $x = 6$ (f) $\frac{n^3}{9} - 1$ for $n = 3$

(g) $4xy - 5x$ for $x = 3$ and $y = 2$ (h) $5ab^2 - 4a^2$ for $a = 5$ and $b = 2$

(i) $(5x)^2$ for $x = 3$ (j) $(x^2 + 2x)^2$ for $x = 3$

(k) $5y - y^2 + 4y^2$ for $y = 8$ (l) $4x - x^3 - 6x^2$ for $x = 3$

(m) $8p + p^2 - p^4$ for $p = \frac{1}{2}$ (n) $-3x^3 + \frac{1}{3}x^3 + \frac{1}{2}x^5$ for $x = 2$

(o) $12r + 6r^2 - 2r^3$ for $r = -2$ (p) $-6p - \frac{6}{p} - (-2p)^2$ for $p = 5$

(q) $5xy + 2xy^2 - 2x^2y^2$ for $x = 3$ and $y = -2$

(r) $3x + 4y^2$ for $x = -3$ and $y = -4$

(s) $(2x)^2 - y^3$ for $x = -\frac{1}{2}$ and $y = \frac{1}{4}$

(t) if $c = -4$ find $8c + 27$ $3c^2 - 4c$ $\frac{c^3}{16}$

(u) if $d = -6$ find $16 - 3d$ $2d^2 + \frac{d}{2}$ $\frac{2}{7}(d^2 + 6)$

2.4 Operations on algebraic fractions

1. Multiplication and division

Treat algebraic fractions in the same way as arithmetic fractions. Apply the same rules.

Multiplication

To multiply algebraic fractions:
1. Factorise the numerator and denominator where possible.
2. Cancel like terms between the numerator and denominator.
3. Multiply out the remaining terms.

Division

To divide algebraic fractions:
1. Factorise the numerator and denominator where possible.
2. *Invert the denominator* term (as with number fractions) and change the sign to *multiply*.
3. Proceed as for multiplication of fractions.

Example 1

Simplify $\frac{4a}{b} \times \frac{b^2}{a}$

Worked solution

$\frac{4a}{b} \times \frac{b^2}{a}$ (Cancel like terms first)

$= 4b$

Example 2

Simplify $\dfrac{\frac{5xy}{6}}{\frac{y^2}{18x^3}}$

Worked solution

$\dfrac{\frac{5xy}{6}}{\frac{y^2}{18x^3}}$

$= \frac{5xy}{6} \times \frac{18x^3}{y^2}$ (Invert, multiply and cancel like terms first)

$= \frac{15x^4}{y}$

Example 3

Simplify $\frac{(4x^2 - 9y^2)}{(4x + 6y)}$

Worked solution

$\frac{(4x^2 - 9y^2)}{(4x + 6y)}$ (Factorise numerator and denominator)

$= \frac{(2x + 3y)(2x - 3y)}{2(2x + 3y)}$ (Cancel like terms)

$= \frac{(2x - 3y)}{2}$

$= \frac{1}{2}(2x - 3y)$

---

2. Addition and subtraction

Apply the same rules to algebraic fractions as for arithmetic fractions.
1. Find the LCM of the denominators.
2. Convert each fraction's numerator to the same LCM (i.e. multiply the answer from the division of each denominator into the LCM by its respective numerator).
3. Simplify the numerator by collecting like terms.
4. Cancel any like terms between the numerator and the denominator.

Example 1

Calculate $\frac{a}{4} + \frac{b}{12}$

Worked solution

$\frac{a}{4} + \frac{b}{12}$ (LCM = 12)

$= \frac{3a}{12} + \frac{b}{12}$ (Convert each fraction's numerator to the same LCM)

$= \frac{(3a + b)}{12}$ (Add the numerators)

Example 2

Calculate $\frac{8}{x} - \frac{5}{y}$

Worked solution

$\frac{8}{x} - \frac{5}{y}$ (LCM = xy)

$= \frac{8y}{xy} - \frac{5x}{xy}$ (Convert each fraction's numerator to the same LCM)

$= \frac{(8y - 5x)}{xy}$ (Add the numerators)

Exercise 24

1. Simplify the following:

   (a) $\frac{x}{4} + \frac{5y}{8x}$ (b) $\frac{3}{2y} - \frac{8y}{5z}$ (c) $\frac{7a}{6b} - \frac{9b}{4c}$

   (d) $\frac{10}{3t} + \frac{u^2}{5v}$ (e) $\frac{a}{3} + a - \frac{a}{5}$ (f) $\frac{x}{3} + \frac{x}{4} - \frac{x}{6}$

   (g) $\frac{7}{6ab} + \frac{9a}{8b^2}$ (h) $\frac{1}{m^2} + \frac{1}{2m} - \frac{7m}{5}$

2. Simplify the following:

   (a) $\frac{6m}{n} \times \frac{n^2}{2m^2}$ (b) $\frac{6v^2}{u^2} \times \frac{u^3}{2v}$

   (c) $\frac{28x^2}{15y^2} \times \frac{5y}{7x^3}$ (d) $\frac{p^3}{24q^2} \times \frac{40q}{p} \times \frac{6}{q^3}$

   (e) $\frac{x}{y^2} \div \frac{6}{y^5}$ (f) $\frac{2a}{3b} \div \frac{8a^3}{2}$

(g) $\dfrac{6x^2y}{7y^4} \div \dfrac{5x^3}{7y^3}$

(h) $\dfrac{3}{a} \div \dfrac{4ba^3}{5} \times \dfrac{15a^3}{2b^3}$

(i) $\left[\dfrac{(2a + 6b)}{5c}\right] \times \left[\dfrac{15c^2}{(a + 3b)}\right]$

(j) $\dfrac{(4a + 16b)}{8} \div \dfrac{(2a + 8b)}{6}$

(k) $\left[\dfrac{(4m - 8n)}{30n^2}\right] \times \left[\dfrac{6n}{(m - 2n)}\right]$

(l) $\dfrac{(p^2 - q^2)}{24r^2} \times \dfrac{9r^3}{(p + q)}$

(m) $\dfrac{(10x - 5y)}{48xy^2} \div \dfrac{(4x^2 - y^2)}{16x^3y}$

(n) $\dfrac{20x^3}{(15x + 5y)} \times \dfrac{(6x + 2y)}{32x^2}$

2.5 Indices

The rules for indices in algebra are identical to the rules for indices in arithmetic,

e.g.   in arithmetic: $4 \times 4 \times 4 = 4^3$ in algebra: $a \times a \times a = a^3$

Remember: $x^0 = 1$ (Recall that any base number to the power of zero = 1)

Rule 1 When *multiplying* like bases, *add the powers*.

Rule 2 When *dividing* like bases, *subtract the powers*.

Rule 3 When a *base with a power is raised to a power, multiply the powers* together.

Rule 4 $(ab)^x = a^x b^x$ or $\left(\dfrac{a}{b}\right)^x = \dfrac{a^x}{b^x}$ (distributive law of indices).

Rule 5 $a^x b^x = (ab)^x$ or $\dfrac{a^x}{b^x} = \left(\dfrac{a}{b}\right)^x$ (grouping bases with the same powers).

Rule 6 $a^{-x} = \dfrac{1}{a^x}$ (converting negative powers to positive powers).

Example 1	Example 2	Example 3
$a^2 \times a^4 = a^{2+4} = a^6$	$\dfrac{a^5}{a^2} = a^{(5-2)} = a^3$	$(a^3)^5 = a^{(3 \times 5)} = a^{15}$

Exercise 25

1. Simplify.

(a) $q^3 \times q^6$

(b) $3^3 \times 3^2$

(c) $y^{-3} \times y^{-5}$

(d) $3a^3 \times 4a^{-1}$

(e) $7m^4 \times 8m^{-4}$

(f) $m^6p \times m^{-4}p^{-2}$

(g) $\dfrac{z^{12}}{z^8}$

(h) $\dfrac{m^8}{m^5}$

(i) $\dfrac{5^{16}}{5^{14}}$

(j) $\dfrac{15p^7}{3p^4}$

(k) $\dfrac{72r^{10}}{8r^6}$

(l) $\dfrac{7a^{12}}{35a^7}$

(m) $\dfrac{8b^{18}}{24b^{18}}$

(n) $\dfrac{xy^6}{x^6y}$

(o) $\dfrac{a^8b^{-2}}{a^3b^6}$

(p) $(a^3)^6$

(q) $(x^2)^{(-3)}$

(r) $(3y^3)^3$

(s) $(1^8)^4$

(t) $(10^2)^3$

(u) $(a^3b^5)^3$

(v) $(2n^2)^5$

(w) $\left(\dfrac{1}{2}y^2\right)^4$

(x) $\left(\dfrac{2}{3}q^5\right)^4$

2.6 Algebraic equations and functions

1. Changing the subject of a formula

1. Collect all terms containing the subject, x, on the left-hand side (LHS) of the equation and all terms not containing the subject, x, on the right-hand side (RHS) of the equation.

   Note: Do not forget to *change the sign* of the terms when moving them across the equal sign.

2. Simplify both sides of the equation – reduce the left-hand side to a single term where the subject, x, is a factor of the left-hand term. This may require factorising the left-hand-side terms.

3. Make x the subject of the formula by *dividing all terms* (both LHS and RHS) by the *coefficient* of x (i.e. the other factor of the LHS term).

Example 1	**Example 2**
Make x the subject of the formula $3x - 2y = 6$.	Make x the subject of the formula $3y + 2 = \frac{6}{x}$.
Worked solution	**Worked solution**
$3x - 2y = 6$ $3x = +2y + 6$ (Move $-2y$ to the RHS) $x = \frac{2y}{3} + 2$ (Divide both sides by 3)	$3y + 2 = \frac{6}{x}$ (Multiply both sides by x) $3xy + 2x = 6$ (Factorise the LHS) $x(3y + 2) = 6$ (Divide both sides by $(3y + 2)$) $x = \dfrac{6}{(3y + 2)}$

2. Solving simple equations

A **simple equation** has only *one unknown* that needs to be solved for.

The method for solving simple equations is exactly the same as for **changing the subject of a formula** (see 2.6 part 1 above).

Example 1	**Example 2**
Solve for x: $3x + 4 = -5$.	Solve for x: $\frac{16}{x} - 3 = 5$.
Worked solution	**Worked solution**
$3x + 4 = -5$ $\quad 3x = -5 - 4$ $\quad 3x = -9$ $\quad\; x = -\frac{9}{3}$ $\therefore x = -3$	$\frac{16}{x} - 3 = 5$ (Multiply through by x) $16 - 3x = 5x$ $\qquad 16 = 5x + 3x$ (Move all x-terms to the RHS) $\qquad 16 = 8x$ $\qquad \therefore x = \frac{16}{8}$ $\qquad \therefore x = 2$

Example 3

Solve for m: $\dfrac{m}{8} = \dfrac{1}{24}$.

Worked solution

$\dfrac{m}{8} = \dfrac{1}{24}$ (Multiply both sides by 8)

$m = 8 \times \dfrac{1}{24}$ (Simplify RHS)

$m = \dfrac{1}{3}$

3. Solving for powers of x

When the *unknown term*, x, is a *power*, the key to solving the equation is to construct a *single base* of the *same number* on *each side of the equation*.

To solve for powers of x:
1. Make the bases the same on both sides of the equation.

> **Note:** There must be *only one base number* on each side of the equation. If the *like* bases are either *added* or *subtracted*, *first factorise* (use the common factor) to convert the expression into a *single base* term:
>
> e.g. $4^x + 4^{(x-1)} = 4^x(1 + 4^{-1}) = 4^x\left(\dfrac{5}{4}\right)$

2. Drop the bases and equate the expressions in the powers and solve for x.

Example 1

Solve for x: $2^x = 2^{2x-3}$.

Worked solution

$2^x = 2^{2x-3}$ (Bases are the same)

$x = 2x - 3$ (Drop same bases)

$3 = 2x - x$ (Group like terms)

$x = 3$

Example 2

Solve for x: $9^x = 3^{5x-3}$.

Worked solution

$9^x = 3^{5x-3}$

$(3^2)^x = 3^{5x-3}$ (Reduce to same base)

$3^{2x} = 3^{5x-3}$

$2x = 5x - 3$ (Drop same bases)

$-3x = -3$

$x = 1$

4. Functions

A function shows a relationship between an input value, x, and an output, written as $f(x)$:
e.g. $f(x) = 2x + 3$ is a function.
To evaluate a function, a value is assigned to x and then substituted into the relationship:
e.g. what is the value of $f(x) = 2x + 3$ for $x = 5$? It is written as $f(5) = 2(5) + 3 = 13$.
Thus $f(5) = 13$.
Other examples of functions are: $f(x) = 2x^2 - 3x + 5$; $f(x) = \dfrac{(4x+3)}{x}$; $f(x) = (x + 2)^3 - 6$.

Rule: The x-value is always substituted into the function in the x-position to evaluate the function.

Example

For the function $f(x) = x^2 + 3$, what is the value of (a) $f(5) - f(3)$ and (b) $\frac{f(6)}{f(0)}$?

Worked solution

(a) $f(5) = (5)^2 + 3 = 28$

   $f(3) = (3)^2 + 3 = 12$

   then $f(5) - f(3) = 28 - 12 = 16$

(b) $f(6) = (6)^2 + 3 = 39$

   $f(0) = (0)^2 + 3 = 3$

   then $\frac{f(6)}{f(0)} = \frac{39}{3} = 13$

Exercise 26

1. Make the unknown in square brackets the subject of the formula:

   (a) $3a + 2b = a + 5c$ $[a]$ (b) $-2y - 7z = 5z + 4y$ $[y]$

   (c) $m^2y = m + 1 + y$ $[y]$ (d) $y + 2 = \frac{5}{x} - 3$ $[x]$

2. Solve each of the following equations:

   (a) $c + 18 = 29$ (b) $-18 = -p - 24$ (c) $3y - 5 = 10$

   (d) $\frac{a}{2} = -5$ (e) $\frac{c}{4} = 2\frac{1}{2}$ (f) $\frac{8}{x} = \frac{2}{3}$

   (g) $\frac{2x}{3} = \frac{5}{6}$ (h) $1\frac{1}{5}a - 1 = 5$ (i) $3(2x - 1) = 6$

   (j) $2(3x - 5) = 6$ (k) $6x + 5 = 3x + 14$ (l) $12x - 16 = 7x - 1$

   (m) $2(2x - 1) = x + 4$ (n) $x + 2(x + 4) = -4$ (o) $6(x - 5) = 5(x - 4)$

   (p) $\frac{(x + 2)}{5} = \frac{(x + 4)}{7}$ (q) $\frac{(x - 4)}{3} = \frac{(x - 11)}{10}$ (r) $\frac{(3x + 2)}{2} = \frac{(6x + 11)}{5}$

   (s) $3x - 4 = 8$ (t) $-3x + 4 = 13$ (u) $-3p - 21 = -12$

   (v) $7 - x = 5$ (w) $x - 4 = -6$ (x) $3 - x = -9$

   (y) $5c - 12 = -2$ (z) $\frac{a}{4} + 2 = \frac{9}{2}a$

3. Solve for x:

   (a) $2^{2x} = 8^{(2x - 4)}$ (b) $9^{(x - 2)} = 27^{(x - 3)}$

   (c) $25^{(x - 3)} = 5^{(x - 4)}$ (d) $16^x = 8^{(2x - 4)}$

4. Find n such that $5^{21} \times 4^{11} = 2 \times 10^n$

5. Find m such that $\left(\frac{1}{5}\right)^m \times 4^{17} = \frac{1}{2} \times 10^{35}$

6. If $2^x 3^y = 288$, where x and y are positive integers, what is the value of $2^{(x - 1)}3^{(y - 2)}$?

7. Find x if $3^x - 3^{(x - 1)} = 162$

8. Find r if $2^r - 2^{(r - 2)} = 3 \times 2^{13}$

9. $12^x 4^{(2x + 1)} = 2^k 3^2$. What is the value of k?

10. For which of the following functions does $f(a) - f(b) = f(a - b)$ for all values of a and b?

   (a) $f(x) = x^2$ (b) $f(x) = \frac{x}{2}$ (c) $f(x) = x + 5$ (d) $f(x) = 2x - 1$ (e) $f(x) = |x|$

   Hint: Replace the unknowns, a and b, with simple integer numbers.

11. For which *one* of the following functions is $f(x) = f(1 - x)$ for all x?

(a) $f(x) = 1 - x$ (b) $f(x) = 1 - x^2$ (c) $f(x) = x^2 - (1 - x)^2$ (d) $f(x) = x^2(1 - x)^2$

(e) $f(x) = \dfrac{x}{(1 - x)}$

Hint: Substitute simple integer numbers for x.

2.7 Factorising

To factorise an expression means to find at least two factors (numbers or terms) which, when multiplied together, equal the expression. This is the same as in arithmetic (e.g. 24 can be factorised into 4 and 6, since 4 × 6 = 24).

1. Common factor

A **common factor** is a value which is *common to all terms* in an expression.

Example 1	**Example 2**
Factorise $2x^2 + 3xy - x$	Factorise $9ab - 3bc + 6bd$
Worked solution	**Worked solution**
$2x^2 + 3xy - x$ (Common factor is x) $= (x)(2x + 3y - 1)$	$9ab - 3bc + 6bd$ (Common factor is $3b$) $= (3b)(3a - c + 2d)$

Note: Each bracketed term is a *factor* of the expression.

2. Difference between two squares

The expression $(x^2 - y^2)$ is known as the *difference between two squares*.
The *factors* of this expression are $(x + y)$ and $(x - y)$, since $(x + y)(x - y) = x^2 - y^2$.

Example 1	**Example 2**
Factorise $x^2 - 81$	Factorise $16x^2 - 25y^2$
Worked solution	**Worked solution**
$x^2 - 81$ $= x^2 - 9^2$ $= (x + 9)(x - 9)$	$16x^2 - 25y^2$ $= (4x)^2 - (5y)^2$ $= (4x + 5y)(4x - 5y)$

3. Using the difference between two squares to simplify arithmetic calculations

Example 1	**Example 2**
Find the value of $51^2 - 49^2$	Find the value of $\left(7\frac{1}{2}\right)^2 - \left(2\frac{1}{2}\right)^2$
Worked solution	**Worked solution**
Since this expression $(51^2 - 49^2)$ is a difference between two squares, factorise as follows: $(51 + 49)(51 - 49) = 100 \times 2 = 200$	Since this expression, $\left(7\frac{1}{2}\right)^2 - \left(2\frac{1}{2}\right)^2$, is a difference between two squares, factorise as follows: $\left(7\frac{1}{2} + 2\frac{1}{2}\right)\left(7\frac{1}{2} - 2\frac{1}{2}\right) = 10 \times 5 = 50$

4. Quadratic expressions

The expression $ax^2 + bx + c$ is a quadratic expression.

To factorise, find two terms which, when multiplied together, equal the quadratic expression. The following method minimises the number of trial-and-error attempts to find the correct pair of factor numbers.

Method (using $3x^2 + 11x + 6$ as an example)

1. Multiply the coefficient of x^2 (i.e. 3) and the constant (i.e. 6) together (i.e. $3 \times 6 = 18$).

2. Find two factors of this product (i.e. 18) such that, when either *added together*, or *subtracted from each other*, the answer is equal to the *middle-term coefficient* (i.e. choose the factors 2 and 9, since $2 \times 9 = 18$, and $2 + 9 = 11$).

3. Once the correct pair of numbers is found, re-write the original quadratic expression but split the middle term (i.e. $11x$) into these two factors (i.e. $3x^2 + 2x + 9x + 6$).

4. Now take out a common factor from the first two terms, and take out a common factor from the last two terms:
   i.e. $3x^2 + 2x + 9x + 6 = x(3x + 2) + 3(3x + 2)$.

5. Finally, take out the bracketed term as a common factor from the expression: i.e. $(3x + 2)(x + 3)$.

Note: The sign of the constant term, c, is important:

(a)   If the sign of c is **+**, then the *two factors* at step 2 must be *added together* to give the middle-term coefficient. As a result, the *two brackets* will have the *same sign*, which will be either *both* **+** if the middle term was positive, or *both* **–** if the middle term was negative.

(b)   If the sign of c is **–**, then the *two factors* at step 2 must be *subtracted from each other* to give the middle-term coefficient. As a result, the two brackets will have opposite signs.

> **Example**
> Factorise $5x^2 - 8x - 4$.
>
> **Worked solution**
> The – sign of the constant ($c = -4$) tells us that the two factors at step 2 must be subtracted from each other. Also, the two brackets will have opposite signs.
>
> **Step 1**   Find the product of 5 and 4 = 20.
> **Step 2**   Choose factors 10 and 2. Since the sign of the constant (–4) is negative, these two factors of 20 must be subtracted from each other to give a value of –8,
> i.e. $10 \times 2 = 20$ and $-10 + 2 = -8$

Step 3 Re-write, splitting the middle term into the two factors –10 and +2:
$$5x^2 - 10x + 2x - 4$$
Step 4 Remove the common factor $5x$ from the first two terms, and the common factor 2 from the last two terms:
$$5x(x - 2) + 2(x - 2)$$
Step 5 Remove the common factor $(x - 2)$ from each term of the expression, giving $(x - 2)(5x + 2)$
This is the factorised form of the quadratic expression.

5. Solving a quadratic equation

When a quadratic expression is set **equal to zero**, it is called a **quadratic equation**, e.g. $5x^2 - 8x - 4 = 0$.

Roots are the *solution values* of a quadratic equation. These are the values of x that make the equation true (i.e. LHS = RHS).

The *graph* of a quadratic equation is called a **parabola** and the points where the graph *intersects the x-axis* are called the *roots* (or the solution values) of the equation.

To find the roots of a quadratic equation:
1. Factorise the quadratic expression on the left-hand side of the equation.
2. Set each bracket (or factor) equal to 0. This results in two simple equations. (Either factor must be 0 to make the LHS = RHS of the quadratic equation.)
3. Solve each simple equation for the value of x that makes each factor equal to zero. These x-values are the roots of the quadratic equation.

Example
Find the roots (solution values) of $5x^2 - 8x - 4 = 0$.

Worked solution

Step 1 Factorise $(5x^2 - 8x - 4)$:
i.e. $(x - 2)(5x + 2)$ (see the previous example)
Step 2 Now either $(x - 2) = 0$ or $(5x + 2) = 0$ for the equation to be true.
Step 3 Solve for x: $x = 2$ or $x = -\frac{2}{5}$.
(These are the roots of the quadratic equation.)

6. Finding the quadratic equation when given the roots

If the roots of a quadratic equation are given, the quadratic equation can be found by reversing the above three steps.

Example

If $x = 3$ and $x = -\frac{3}{4}$ are the roots of a quadratic equation, find the equation.

Worked solution

Given that $x = 3$ and $x = -\frac{3}{4}$, then $(x - 3) = 0$ and $\left(x + \frac{3}{4}\right) = 0$.

Multiply $\left(x + \frac{3}{4}\right) = 0$ by 4 to remove the fraction, i.e. $(4x + 3) = 0$.

Each term, $(x - 3)$ and $(4x + 3)$, is a *factor* of the quadratic equation.

Therefore $(x - 3)(4x + 3) = 0$ is the *factorised form* of the quadratic equation.

Multiply out the brackets, giving $4x^2 - 9x - 9 = 0$.

The *quadratic equation* is then $y = 4x^2 - 9x - 9$.

Exercise 27

1. Factorise (common factor):
    - (a) $6m + 2mn - 18m^3$
    - (b) $10pqr + 6p^2qr^2$
    - (c) $-8x^3y^2 - 12x^2 - 16x^4y^3$
    - (d) $6a^3b^2c - 12ab^2 + 6a^2bc$

2. Factorise (difference between two squares):
    - (a) $4x^2 - 25$
    - (b) $x^2 - \frac{1}{4}$
    - (c) $b^2 - \frac{25}{36}$
    - (d) $25m^2 - 16$
    - (e) $9a^2 - \frac{1}{16}$
    - (f) $49x^2 - 9y^2$

3. Find the value of:
    - (a) $25^2 - 15^2$
    - (b) $38^2 - 2^2$
    - (c) $29^2 - 11^2$
    - (d) $(0.6)^2 - (0.4)^2$
    - (e) $(0.9)^2 - (0.1)^2$
    - (f) $(6.5)^2 - (3.5)^2$
    - (g) $\left(6\frac{2}{3}\right)^2 - \left(2\frac{1}{3}\right)^2$
    - (h) $\left(5\frac{7}{8}\right)^2 - \left(4\frac{1}{8}\right)^2$
    - (i) $(1)^2 - (0.6)^2$

4. Factorise (quadratic expressions):
    - (a) $y^2 + 10y + 16$
    - (b) $a^2 - 5a - 24$
    - (c) $6x^2 + 13x - 5$
    - (d) $2r^2 - 19r + 42$
    - (e) $t^2 - 11t + 18$
    - (f) $10a^2 + 7a - 12$
    - (g) $15x^2 - 17x - 4$
    - (h) $2y^2 - y - 21$
    - (i) $6c^2 - 13c - 5$
    - (j) $6x^2 + x - 15$
    - (k) $2x^2 - x - 1$
    - (l) $12x^2 - 7x - 12$
    - (m) $x^2 - 5x - 14$
    - (n) $8x^2 - 20x + 8$

5. For each quadratic expression in question 4 (a) to (n), form a quadratic equation. Solve for the roots of each quadratic equation.

6. Which one of these equations has a root in common with $x^2 - 6x + 5 = 0$?
    - (a) $x^2 + 1 = 0$
    - (b) $x^2 - x - 2 = 0$
    - (c) $x^2 - 10x - 5 = 0$
    - (d) $2x^2 - 2 = 0$
    - (e) $x^2 - 2x - 3 = 0$

7. Find the quadratic equation that has the following two roots:

   (a) $x = 2$ and $x = -3$ 　　　　　(b) $x = \frac{1}{4}$ and $x = \frac{1}{2}$

   (c) $x = -4$ and $x = \frac{1}{2}$ 　　　　(d) $x = -\frac{3}{5}$ and $x = 2$

   (e) $x = 5$ and $x = 2$ 　　　　　(f) $x = -\frac{2}{3}$ and $x = -\frac{3}{5}$

   (g) $x = -\frac{3}{2}$ and $\frac{4}{5}$

8. If 4 is one of the roots of the quadratic equation $x^2 + 3x + (k - 10) = 0$, what is the other root?

9. If $(t - 8)$ is a factor of $t^2 - kt - 48$, what is the value of k?

2.8　Graphs and equations (straight line and parabola)

1. Terms: Cartesian graph; axes; quadrant and coordinates

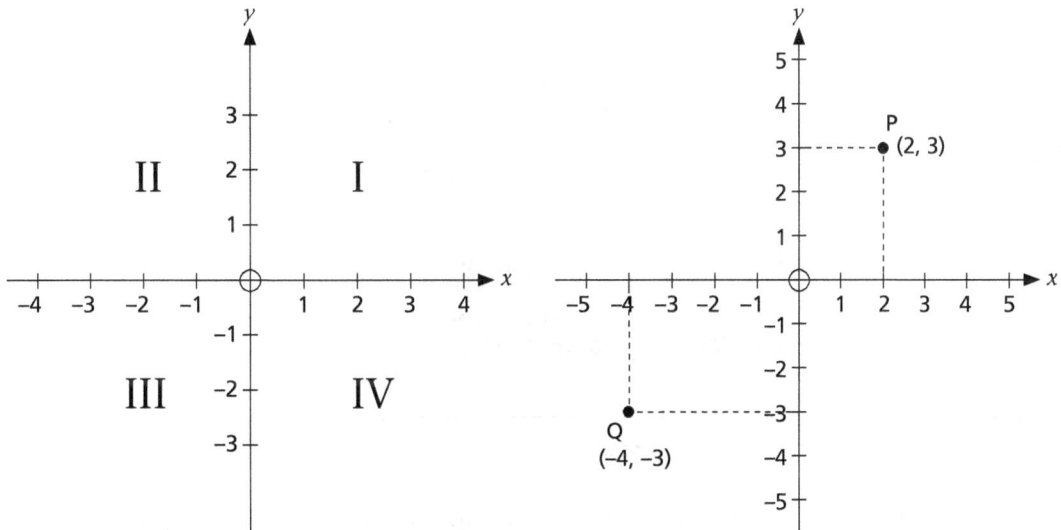

A Cartesian graph is a graph with a *horizontal axis* (*x*-axis) and a *vertical axis* (*y*-axis). Where these **axes** intersect, is the origin.

The *x*- and *y*-axes make four **quadrants** in a graph.
In quadrant I: 　　　x is + and y is +, 　e.g. (+3, +4)
In quadrant II: 　　x is – and y is +, 　e.g. (–2, +5)
In quadrant III: 　　x is – and y is –, 　e.g. (–1, –4)
In quadrant IV: 　　x is + and y is –, 　e.g. (+5, –2)

A **coordinate** is a point on a Cartesian graph. It is labeled *(x, y)*.
The *origin* has the coordinate (0, 0).
Consider the coordinate *P* at (2, 3) in the above graph.
Its *horizontal* distance along the *x*-axis from the origin is 2 units;
its *vertical* distance along the *y*-axis from the origin is 3 units.

---

2. Plotting a straight line graph

The *standard form* of a straight line graph is represented by the equation

$$y = mx + c$$ m = gradient (slope) of the line
c = y-intercept

A **gradient** is defined as $m = \dfrac{\text{vertical distance}}{\text{horizontal distance}}$

The *sign* of the *gradient* indicates the *direction* of a straight line graph.

- If $m > 0$ $\left(\text{e.g. } m = \frac{3}{5}\right)$ it is a *positive-sloped* graph. It runs from bottom left to top right.
- If $m < 0$ $\left(\text{e.g. } m = -\frac{2}{3}\right)$ it is a *negative-sloped* graph. It runs from top left to bottom right.

Only two coordinates are needed to plot any straight line graph.

Method 1: Gradient/intercept method

1. Re-write the equation in the standard form $y = mx + c$.
2. Plot one coordinate at $(0, c)$. This is the y-intercept.
3. Show the gradient as an improper fraction (if it is an integer, divide it by 1).
4. From the y-intercept coordinate $(0, c)$, move horizontally the distance indicated by the denominator of the gradient; then move vertically the distance indicated by the numerator of the gradient. Move in the direction of the signs of the numerator and denominator. The end point defines the second coordinate.
5. Join the two coordinates to show the straight line graph.

Note: If the straight line equation is in a *non-standard form* ($ax + by = d$, where a, b and d are constants), use these short-cut formulae to find the *gradient* and *y-intercept*:

gradient $m = -\dfrac{a}{b}$

i.e. divide the x-coefficient by the y-coefficient and change the sign of the answer

y-intercept $c = \dfrac{d}{b}$

i.e. divide the constant, d, by the y-coefficient.

Method 2: Substitution approach (using a table)

1. Let $x = 0$, then find y (substitute $x = 0$ into the equation and solve for y).
2. Let $y = 0$, then find x (substitute $y = 0$ into the equation and solve for x).
3. Choose one further x-value and solve for y to confirm calculations. This step is not essential.

4. Plot all three coordinates – they should lie on a straight line.

Steps 1 and 2 find the coordinates where the graph cuts the y-axis and x-axis, respectively.

Example 1

For $5x - 2y = 8$, find the gradient and y-intercept using the formulae $m = -\frac{a}{b}$ and $c = \frac{d}{b}$.

Worked solution

$a = 5$; $b = -2$; $d = 8$

Gradient, $\boldsymbol{m} = -\left(\frac{5}{(-2)}\right) = \frac{5}{2}$ $\qquad$ y-intercept, $\boldsymbol{c} = \left(\frac{8}{(-2)}\right) = -4$

Example 2

Plot the straight line graph of $y - 3x = 5$.

Worked solution

Method 1 (gradient/intercept)

1. Re-write the equation in standard form: $y = 3x + 5$.
2. y-intercept is the first coordinate $= (0, +5)$
3. Gradient $= \frac{3}{1}$
4. To find the second coordinate, from $(0, +5)$, move horizontally $+1$ unit and vertically $+3$ units. This ends at coordinate $(+1, +8)$.

Method 2 (Substitution)

Let $x = 0$, then $y = 5$

Let $y = 0$, then $x = -\frac{5}{3}$

Lastly choose, say, $x = 1$, then $y = 8$

x	0	$\frac{-5}{3}$	1
y	5	0	8

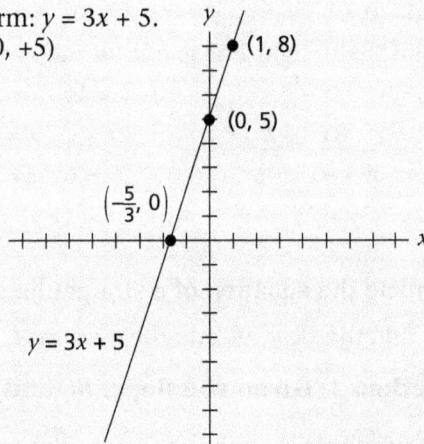

Example 3

Plot the straight line graph of $2y + 3x = 8$.

Worked solution

Method 1: Gradient/intercept

1. Re-write the equation in standard form: $y = \frac{-3}{2}x + 4$.
2. y-intercept is the first coordinate = $(0, +4)$
3. Gradient $= -\frac{3}{2}$
4. To find the second coordinate, from $(0, +4)$, move horizontally $+2$ units and vertically (down) 3 units. This ends at coordinate $(+2, +1)$.

Method 2: Substitution

Complete the table by using substitution:

x	0	$\frac{8}{3}$	1
y	4	0	$\frac{5}{2}$

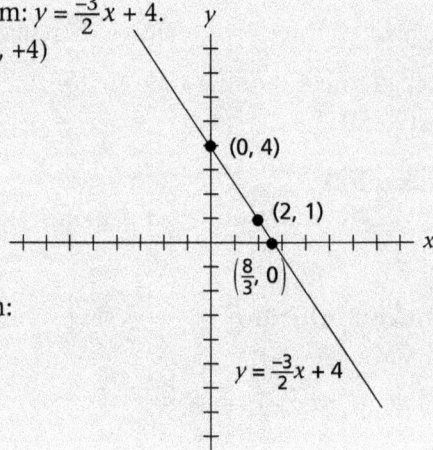

$(0, 4)$

$(2, 1)$

$\left(\frac{8}{3}, 0\right)$

$y = \frac{-3}{2}x + 4$

Exercise 28

1.   In which quadrant does each coordinate lie?

   (a) $(2, -3)$ (b) $(-5, 2)$ (c) $(6, 7)$

   (d) $(0, -1)$ (e) $(5, -2)$ (f) $(6, 0)$

   (g) $(0, 0)$ (h) $(0, 8)$ (i) $\left(-7, -\frac{1}{2}\right)$

2.   Plot the straight line graph for each of the following:

   (a) $y = 2x + 4$ (b) $y = 4x - 3$

   (c) $y + 3x = -2$ (d) $2x - y - 5 = 0$

   (e) $3y - 9x = -6$ (f) $2y = 6 - 5x$

   (g) $-3y - 2x = -9$ (h) $5x + 4y = -6$

3. Finding the equation of a straight line $y = mx + c$

To find the *slope*, m, and the *y-intercept*, c, use these methods:

Method 1: Given the slope, m, and a coordinate (x_1, y_1)

1. Substitute the coordinate into the equation and solve for c.
2. Since the slope, m, is given, the equation can be written as $y = mx + c$.

Example 1

Find the equation with a slope of 2 that passes through the coordinate (3, 1).

Worked solution

Since $m = 2$, the equation can be written as $y = 2x + c$.
Now substitute $x = 3$ and $y = 1$ into the equation $y = 2x + c$ and solve for c.

$$1 = 2(3) + c$$
$$1 - 6 = c$$
$$\therefore c = -5$$

Thus the straight line equation is $y = 2x - 5$.

Method 2: Using two coordinates (x_1, y_1) and (x_2, y_2) on the line

1. Use a formula to find the slope, m, from the two coordinates.
    The formula for *slope* is:

$$\boxed{\textbf{Slope, } m = \frac{\text{change in } y\text{-values}}{\text{change in } x\text{-values}} = \frac{(y_2 - y_1)}{(x_2 - x_1)}}$$

2. Substitute one of the coordinates (chosen arbitrarily) into the equation with the known slope, m, and solve for c (apply Method 1 above).

Example 2

Find the equation that passes through the coordinates (3, 1) and (2, −1).

Worked solution

Slope, $m = \dfrac{(-1 - 1)}{(2 - 3)}$ (From the slope formula)

$\quad\quad\quad = \dfrac{(-2)}{(-1)} = 2$

The equation can now be written as $y = 2x + c$.
Now substitute $x = 3$ and $y = 1$ into the equation and solve for c:

$$1 = 2(3) + c$$
$$1 - 6 = c$$
$$\text{giving } c = -5$$

Thus the straight line equation is $y = 2x - 5$.

4. Distance of a straight line segment between two coordinates

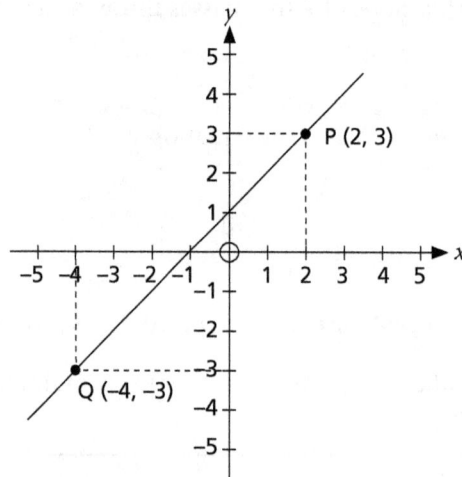

Method Use **Pythagoras' theorem** – which states for a right-angled triangle:

'the square of the hypotenuse equals the sum of the squares of the other two sides'.

In formula terms, distance, d (the length of a hypotenuse):

$$d = \sqrt{(y_2 - y_1)^2 + (x_2 - x_1)^2}$$

where (x_1, y_1) and (x_2, y_2) are two coordinates at each end of the line segment.

Example
Find the length of the line segment between P = (2, 3) and Q = (−4, −3).

Worked solution
$$d = \sqrt{(3 - (-3))^2 + (2 - (-4))^2}$$
$$= \sqrt{(6)^2 + (6)^2}$$
$$= \sqrt{(36 + 36)} = \sqrt{72} = 6\sqrt{2}$$

5. Perpendicular and parallel lines

To check if lines are perpendicular or parallel, always write the equations in the standard straight line form of $y = mx + c$.

(a) Perpendicular lines

Perpendicular lines (i.e. lines at 90° to each other) have reciprocal slopes with opposite signs.

A simple test: Two lines are perpendicular if the *product of their slopes* = −1.

Example

Are the two lines $y = -4x - 3$ and $y = \frac{1}{4}x + 6$ perpendicular to each other?

Worked solution

Test for perpendicular lines: given the slopes $m_1 = -4$ and $m_2 = \frac{1}{4}$
Product of slopes: $m_1 \times m_2 = (-4) \times \left(\frac{1}{4}\right) = -1$
Hence the two lines are perpendicular.

(b) Parallel lines

Two lines are parallel when they have the *same slope* (including the sign).

Example

Are the two lines $2y = -4x - 3$ and $-4y - 8x = 10$ parallel to each other?

Worked solution

Check their slopes, but first write the equations in the standard form, $y = mx + c$.

Equation 1: $2y = -4x - 3$, rewritten as $y = -2x - \frac{3}{2}$ $\therefore$ slope $m_1 = -2$

Equation 2: $-4y - 8x = 10$, rewritten as $y = \left(\frac{8}{-4}\right)x - \frac{10}{4}$ $\therefore$ slope $m_2 = -2$

Since both slopes are equal (in magnitude and sign), the two lines are parallel.

Exercise 29

1. Find the equation for each of the following straight lines:
    (a) passes through the coordinate (2, 3) and has a slope of 6.
    (b) passes through the coordinate (1, 8) and has a slope of –1.
    (c) passes through the coordinate (–2, 7) and has a slope of 0.
    (d) passes through the coordinate $\left(\frac{1}{2}, \frac{2}{3}\right)$ and has a slope of $\frac{1}{4}$.
    (e) passes through the coordinates (–2, 3) and (3, 8).
    (f) passes through the coordinates (1, –1) and (3, 0).
    (g) passes through the coordinates $\left(-\frac{1}{4}, \frac{2}{5}\right)$ and $\left(1, -\frac{1}{2}\right)$.

2. Find the length of the line segment between the coordinates:
    (a) (3, 3) and (9, 11)   (b) (–1, 1) and (2, 5)
    (c) (4, 3) and (16, 8)   (d) (–3, –1) and (5, 5)
    (e) (0, 0) and (–4, –2)   (f) (0, –4) and (3, 0)

3. Which of the following pairs of lines are either perpendicular or parallel to each other?
    (a) $y = 2 + x$ and $y = 3 - x$
    (b) $2x + 3y = 6$ and $2x + 3y = 8$
    (c) $3x - 4y = 12$ and $4x + 3y = 7$
    (d) $x - y = 4$ and $-6 + 2y = 2x$

(e) $y = 4 + 3x$ and $x = 6 + 3y$

(f) $2x - 5y = 7$ and $10x + 4y = 20$

(g) $2x + 3y = 8$ and $9x - 6y = 14$

4.   Find the equation of the line that is perpendicular to the line $2x + y = 5$ and passes through the coordinate $(2, -1)$.

6. Parabola

The standard form of a quadratic equation is: $y = ax^2 + bx + c$

The graph of a quadratic equation is called a **parabola**. An example is shown below.

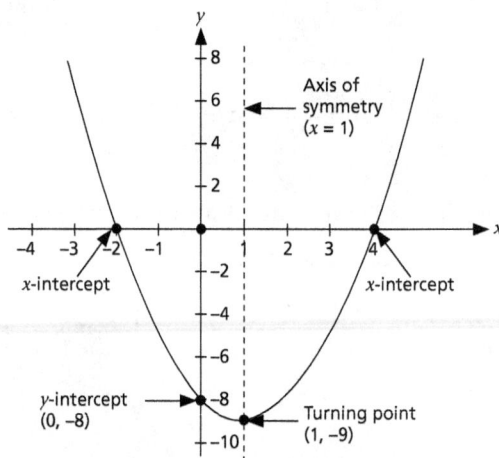

To graph a parabola, you will need to find the *y-intercept*, the *x-intercepts*, which are called the roots, the *axis of symmetry* and the *turning point*.

The following example illustrates these terms.

Example

For $y = 2x^2 + x - 3$ $a = 2$; $b = +1$; $c = -3$

Worked solution

(a) **y-intercept = c** $c = -3$
     Plot the **coordinate $(0, c)$**   e.g. $(0, -3)$

(b) **x-axis intercepts** are called the **roots** (or solution) of a quadratic equation

     To find the roots: *factorise* and solve the equation $ax^2 + bx + c = 0$ (see section 2.6)

     $2x^2 + x - 3 = 0$ $(2x + 3)(x - 1) = 0$ giving $x = -\frac{3}{2}$ or $x = +1$

     Plot the coordinates for the two x-intercepts $\left(-\frac{3}{2}, 0\right)$ and $(1, 0)$.

Note: If the equation cannot be factorised, use the general formula to find the roots:

$$x = \frac{-b \pm \sqrt{b^2 - 4ac}}{2a}$$

(c) **axis of symmetry:** This is a vertical line at $x = -\frac{b}{2a}$,

e.g. $x = -\frac{1}{2(2)} = -\frac{1}{4}$

The parabola is a mirror image about this vertical line at $x = -\frac{1}{4}$.
Note that the axis of symmetry is the midpoint between the two roots on the x-axis.

(d) **turning point:** The parabola turns at the coordinate $\left(-\frac{b}{2a}, c - \frac{b^2}{4a}\right)$

e.g. $-\frac{b}{2a} = -\frac{1}{4}$ and $\left(c - \frac{b^2}{4a}\right) = \left(-3 - \frac{(1^2)}{4(2)}\right)$

$$= \left(-3 - \frac{1}{8}\right)$$

$$= -3\frac{1}{8}$$

Thus the turning point coordinate $= \left(-\frac{1}{4}, -3\frac{1}{8}\right)$

Note: The y-value of the coordinate is commonly found by substituting the axis of symmetry value for x into the standard equation $y = ax^2 + bx + c$,

e.g. $x = -\frac{1}{4}$, then $y = 2\left(-\frac{1}{4}\right)^2 + \left(-\frac{1}{4}\right) - 3 = -\frac{25}{8} = -3\frac{1}{8}$

Exercise 30

1. For each quadratic equation:
    (i) Find the y-intercept, the roots, the axis of symmetry and the turning point.
    (ii) Sketch a parabola graph.

    (a) $y = x^2 + 10x + 16$ (b) $y = x^2 - 5x - 24$ (c) $y = 6x^2 + 13x - 5$
    (d) $y = 2x^2 - 19x + 42$ (e) $y = x^2 - 11x + 18$ (f) $y = 10x^2 + 7x - 12$
    (g) $y = 15x^2 - 17x - 4$ (h) $y = 2x^2 - x - 21$ (i) $y = 6x^2 - 13x - 5$
    (j) $y = 4x^2 - 6x - 4$ (k) $y = 2x^2 - x - 1$ (l) $y = 2x^2 - 5x - 25$
    (m) $y = x^2 - 5x - 14$

2.9   Solving simultaneous equations

For any two straight line equations, the purpose is to find values of x and y that satisfy both equations simultaneously.

Graphically, the (x, y) value is the coordinate where the two straight line graphs intersect.

Note: There is no solution if the two straight lines are parallel to each other.

Method 1: Elimination of one unknown

1. Choose an unknown to eliminate.
2. Make the coefficient of this unknown the same in both equations by multiplying either one or both equations by a constant (a different constant can be used for each equation).
3. Now either add or subtract the two equations to eliminate the chosen unknown.
4. Solve the resulting single equation for the remaining unknown.
5. Substitute the value from (4) into any one of the equations and solve for the other unknown.

Example

Solve for x and y simultaneously:

$5x + 3y = 21$ (Equation [1])
$2x + y = 8$ (Equation [2])

Worked solution

1. Choose to eliminate y.
2. Multiply equation [2] by 3 to make all the coefficients of y the same.
$$(3 \times 2)x + 3 \times y = 3 \times 8$$
   giving $6x + 3y = 24$ (Equation [2])
3. Subtract equation [2] from equation [1] to eliminate y:
$$5x + 3y = 21 \qquad \text{(Equation [1])}$$
$$6x + 3y = 24 \qquad \text{(Equation [2])}$$
4. Resulting in $-x = -3$ Thus $x = 3$
5. Now substitute $x = 3$ into equation [1] (chosen arbitrarily) and solve for y.
$$5(3) + 3y = 21$$
$$15 + 3y = 21$$
$$3y = 21 - 15 = 6$$
$$\text{Thus } y = 2$$

The simultaneous solution is: $x = 3$ and $y = 2$

(Check by substituting (3, 2) into equation [2]:
LHS = 2(3) + 2 = 8; RHS = 8)

Method 2: Substitution of one equation into the other equation

1. Make one of the unknowns from one equation (either x or y) the subject of the formula.
2. Substitute this formula for the unknown into the other equation.
3. Solve the other equation for the remaining unknown.
4. Substitute the value from (3) into the equation chosen in (1) and solve for the other unknown.

Example

Solve for x and y simultaneously:

$5x + 3y = 21$ (Equation [1])
$2x + y = 8$ (Equation [2])

Worked solution

1. From equation [2]:
    $y = -2x + 8$ (making y the subject of the formula)
2. Substitute $y = -2x + 8$ into equation [1], giving $5x + 3(-2x + 8) = 21$
3. Solve equation [1] for x:
    $5x - 6x + 24 = 21$ (remove brackets)
    $-x = 21 - 24$
    $x = 3$
4. Now substitute $x = 3$ into equation [1] (chosen arbitrarily) and solve for y.
    $5(3) + 3y = 21$
    $15 + 3y = 21$
    $3y = 21 - 15 = 6$
    Thus $y = 2$

The simultaneous solution is: $x = 3$ and $y = 2$
(Check by substituting into equation [2]: LHS = 2(3) + 2 = 8; RHS = 8)

Exercise 31

1. Solve the following pairs of simultaneous equations:

    (a) $5x + y = 28$ (b) $9a + b = 13$ (c) $m + 8n = 26$
    $x - y = 2$ $3a + b = 7$ $m + n = 5$

    (d) $3x + y = 13$ (e) $3m + n = 10$ (f) $3p + 4q = 25$
    $5x - 2y = 7$ $5m - 2n = 2$ $p + 2q = 11$

    (g) $5b + 4c = 23$ (h) $3x + 2y = 13$ (i) $5x + 3y = 11$
    $3b + c = 11$ $5x - 3y = 9$ $7x - 2y = 3$

    (j) $4t + 3u = 14$ (k) $5a + 3b = 25$ (l) $2p + 5q = 16$
    $5t - 2u = 6$ $3a + 2b = 16$ $3p - 2q = 5$

    (m) $3x + 4y = 12$
    $2x - y = 8$

Expressing word problems as simple (single) or simultaneous (two) equations

Method

1. Decide how many unknowns there are in the question (i.e. either one or two unknowns).
    *   If there is only *one unknown,* the problem will be written in terms of only *one simple equation* (e.g. $3x - 5 = 18$).
    *   If there are *two unknowns,* the problem will be written in terms of a *pair of simultaneous equations* (e.g. $4x + 2y = 12$ and $x - 3y = 9$).

Tip: The actual question in a word problem is a good guide to decide what is to be found (i.e. how to define the unknown quantities).

2. Use letters to define the quantity or quantities which have to be found,
    e.g. let x = number of kilometres driven; let y = number of tins filled per hour.
3. From information given in the question, write either one or at most two equations containing the defined x- and y-unknowns to represent the wording of the question.
4. Solve either the single equation for the one unknown (x), or the pair of simultaneous equations for the two unknowns (x, y).
5. Check that the answer makes sense in terms of the facts in the question.

Example 1

After a car journey, a motorist found that only a third of the petrol that he started with is left in the tank. At a garage, he puts in 10 litres; then he finds that there are 17 litres in the tank altogether. How many litres were in the tank at the start of the journey?

Worked solution

The actual question asks for '*the number of litres at the start of the journey*'.

This is a **single equation problem** as there is only *one unknown* to be found. So let x = the number of litres at the start of the journey.

From the information given:
(a) The amount of petrol left after the journey but before the filling up = $\frac{1}{3}x$.
(b) Then 10 litres are added to the tank,
     i.e. $\frac{1}{3}x + 10$ = amount of petrol after filling up.
(c) The tank then holds 17 litres, thus $\frac{1}{3}x + 10 = 17$.
(d) Solving $\frac{1}{3}x + 10 = 17$ gives $\frac{1}{3}x = 17 - 10 = 7$ therefore $x = 21$ litres.
(e) Check the answer:
     If the motorist started with 21 litres and finished with $\frac{1}{3}$ of this amount, there were $\left(\frac{1}{3} \times 21 \text{ litres}\right) = 7$ litres left before filling up.
     If 10 litres were added, this makes the tank 17 litres full, as stated in the question.
(f) Answer: *The motorist had 21 litres in the tank at the start of the journey.*

Example 2

A factory produces chairs and tables. Each product must go through a cutting and sanding operation. Each chair requires 3 minutes of cutting and each table takes 5 minutes to cut. A total of 220 minutes of cutting time is available each day. Each chair also takes 2 minutes to sand, while each table takes 7 minutes to sand. There is also a total of 220 minutes of sanding available each day. How many chairs and tables can this factory produce each day?

The actual question asks for '*the number of chairs to produce each day*' and '*the number of tables to produce each day*'.

This is a **simultaneous equations problem** as there are two **unknowns** to be found.

Let x = the number of chairs to produce each day, and
let y = the number of tables to produce each day.

From the information given:

(a) Cutting time is limited to 220 minutes per day and each chair requires 3 minutes and each table requires 5 minutes.
$$3x + 5y = 220 \text{ (Equation [1])}$$

(b) Sanding time is limited to 220 minutes per day and each chair requires 2 minutes and each table requires 7 minutes.
$$2x + 7y = 220 \text{ (Equation [2])}$$

(c) Solve the pair of equations simultaneously.
To eliminate x, multiply equation [1] by 2 and equation [2] by 3, giving
$$6x + 10y = 440 \text{ (Equation [1])}$$
$$6x + 21y = 660 \text{ (Equation [2])}$$
Subtract equation [1] from equation [2], giving $11y = 220$ and $y = 20$.
Substitute $y = 20$ into the original equation [1] and solve for x.
$$3x + 5(20) = 220, \text{ giving } 3x = 120 \text{ and } x = 40$$

(d) Check:
Equation [1]: $3(40) + 5(20) = 220$ (correct)
Equation [2]: $2(40) + 7(20) = 220$ (correct)

(e) Answer: $x = 40$ *chairs to produce daily; $y = 20$ tables to produce daily.*

Exercise 32

Solve the following word problems (using algebraic equations).

1. (a) A person spends R36 on chips and cooldrink. A can of cooldrink costs R4 and a packet of chips costs R2. If the person buys 5 cooldrinks, how many packets of chips did she buy?

    (b) A certain number is multiplied by 8. 11 is then subtracted from the product, giving a result of 29. What was the original number?

    (c) A bag of apples is shared evenly between 4 people. One person throws 3 away from her share, because she found them to be overripe. She has 9 left. How many apples were in the bag altogether?

    (d) If 3 is subtracted from a number and the result multiplied by 2, the answer is 3. What was the number?

    (e) If a number is doubled and then 7 is subtracted from this result, the answer is the same as when 5 is added to the number. What was the number?

    (f) If a number is multiplied by 6 and 3 is added to the result, the answer is the same as when 5 is added to the number and the result is then trebled. Find the number.

    (g) The sum of three consecutive integers is 168. Find the largest number.

    (h) The sum of four consecutive odd numbers is 56. What is the value of the largest number?

    (i) The weights of three dogs in kilograms are consecutive even numbers. If the total weight is 72 kilograms, what is the weight of the heaviest dog?

    (j) Find four consecutive integers divisible by 5 and with a sum of 190.

2. (a) A batsman's scores in 4 innings are consecutive numbers which are multiples of 8 and add up to 176 runs. What was his second highest score?

(b) In a weight-lifting competition, the total weight of two lifts by a competitor was 375 kg. If twice the weight of the first lift was 150 kg more than the weight of his second lift, what was the weight of his first lift?

(c) A teacher bought pens for R4 each and pencils for R2.80 each. If she bought a total of 24 pens and pencils for R84, how many pens and pencils did she buy?

(d) Tourist A rented a car for R18 plus R0.10 per kilometer driven. Tourist B also rented a car for R25 plus R0.05 per kilometer driven. If they travelled exactly the same distance and were charged exactly the same for the rentals, how far did they each drive?

(e) If a student had twice the amount of money that he actually has, he would have exactly the amount necessary to buy 4 cooldrinks at R2.25 each and two doughnuts at R1.50 each. How much does he have?

(f) Mike is now 14 years older than Tim. If in 10 years' time Mike will be twice as old as Tim, how old will Mike be in 5 years?

(g) In a certain school, a student either takes a science course or a physics course. 40 more than $\frac{1}{3}$ of all students are taking a science course and the number taking a physics course is $\frac{1}{4}$ of those taking a science course. If $\frac{1}{8}$ of all students in the school are taking physics, how many students are in the school?

(h) The sum of the digits of a two-digit number is 10. This sum is equivalent to 5 times the difference between those two digits. What is the number represented by these two digits if the tens digit is the larger of the two digits?

(i) A person has 9 coins in their pocket made up of R5 and R2 coins. If the value of the R5 coins exceeds the value of the R2 coins by R3, how many R2 coins did this person have in her pocket?

(j) A club sold an average (arithmetic mean) of 9 raffle tickets per member. The average per female member was 12 and the average for males was 7. If there are 40 members in the club, how many female members does the club have?

(k) In an increasing sequence of eight consecutive integers, the sum of the first four integers is 54. What is the sum of the last four integers in the sequence?

(l) Two families go to the theatre. In family A there are two adults and three children. They pay a total of R125 for their tickets. In family B there are three adults and two children. They pay a total of R150 for their tickets. What will family C, with one adult and three children, pay for their tickets?

(m) In a certain company, the ratio of the number of managers to the number of workers is 5 to 72. If 8 additional workers are hired, the ratio of managers to workers becomes 5 to 74. How many managers does the company have?

(n) Two cars, A and B, travel a combined distance of 440 km in a total time of 9 hours. If car B drives for one hour longer than car A, but at a speed 20 km/h less than car A, how far did car A travel?

(o) A litre (1 000 ml) of a blended fruit juice consists of the natural juice of peaches and pears. Each litre of blended fruit juice must contain 88 ml of vitamin A. Peach juice contains 8% vitamin A, while pear juice contains 10% vitamin A. What is the ratio of peach juice to pear juice in each litre (1 000 ml) of the blended juice?

3. (a) The sum of five consecutive integers, each divisible by 5, is 250. What is the difference between the largest and the smallest of these numbers?

(b) Sue lives x floors above the ground of a high-rise building. It takes her 30 seconds per floor to walk down the steps and 2 seconds per floor to ride the elevator. If it takes Sue the same amount of time to walk down the steps to the ground floor as to wait for the elevator for 7 minutes and ride down, on which floor does Sue live?

(c) One hour after Yolanda started walking from X to Y, a distance of 45 km, Bob started walking along the same road from Y to X. If Yolanda's walking rate was 3 km/h and Bob's was 4 km/h, how many kilometres had Bob walked when they met?

(d) Two trains, A and B, started simultaneously from opposite ends of a 100 km route and travelled towards each other on parallel tracks. Train A, travelling at a constant speed, completed the 100 km trip in 5 hours, while train B, travelling at a constant speed, completed the 100 km trip in 3 hours. How many kilometres had train A travelled when it passed train B?

(e) A hiker walked for two days. On the second day she walked 2 hours longer and at an average speed of 1 km per hour faster than on day one. If during the two days she walked a total distance of 64 km and spent a total of 18 hours walking, what was her average speed on day one?

2.10 Inequalities

An **inequality relationship** means that the left side of the relationship is either *less than* or *greater than* the right side (depending on the sign of the relationship).

The solution of an inequality relationship is a *set of values* that satisfies the inequality (e.g. $x \leq 5$). (Its solution set is unlike that of an equation, where the solution is a single unique value, e.g. $x = 5$ only.)

1. Signs

Inequality signs define the *order of numbers* in relation to each other.

The following signs are used:

< means 'is less than'	e.g. $12 < 19$ (12 is less than 19)
≤ means 'is less than or equal to'	e.g. $x \leq 8$ (x is any value smaller than or equal to 8)
> means 'is greater (more) than'	e.g. $21 > 6$ (21 is greater than 6)
≥ means 'is greater (more) than or equal to'	e.g. $y \geq 3$ (y is any value greater than or equal to 3)

Exercise 33

Use the inequality signs, including the = sign, to connect the following:

(a) $3\frac{1}{2}$ and 3.45

(b) $9 + 7$ and $20 - 3$

(c) $4\frac{3}{4}$ and 4.85

(d) $\frac{64}{8}$ and 9

(e) $5 + 21$ and $3.5 + 23.5$

(f) $-12 - 3$ and $-7 - 9$

(g) $8\frac{1}{2}$ and 8.75

(h) −16.25 and −17.15

(i) 21 × 3 and 62.5

(j) $7\frac{2}{7}$ and 7.25

(k) 4.5 and $\frac{27}{6}$

(l) (3 − 9) and (−2 − 3)

(m) $\frac{(-14 + 5)}{3}$ and $\frac{(-12 - 6)}{(-6)}$

(n) $4\frac{3}{4}$ and 4.7

(o) $6\frac{4}{5}$ and 6.8

(p) 5 + 21 and 5.5 + 20.5

(q) $2.9 + 1.6 - 1.1$ and $2\frac{2}{5} + 1\frac{4}{5} - 1\frac{1}{5}$

(r) 6 × 8 and 4 × 12

(s) −7 − 14 + 2 and −10 + 1 − 10

(t) $\frac{144}{12}$ and $\frac{104}{9}$

(u) $\frac{10.8}{0.9}$ and $\frac{165}{15}$

2. Solving inequalities

A **solution set** is the set of all numbers that satisfies the inequality (i.e. makes the statement true).

When solving an *equation,* there are only *one solution* which satisfies it (e.g. $x = 6$). When solving an *inequality,* there is *more than one solution* which satisfies the relationship (e.g. for $x < 6$, all values that are less than 6 are solution values of the inequality $x < 6$).

Inequalities are solved in a similar way to equations, with one major exception:

- If both sides of the inequality are *multiplied* or *divided* by a *negative number, the inequality sign (direction of the inequality) must be reversed.*

Example 1

Find the solution set of all numbers for $6x - 5 < 13$.

Worked solution

Solve the inequality $6x - 5 < 13$ in the same way as solving an equation:
i.e. $6x < 13 + 5$ $6x < 18$ $x < 3$

The *solution set* is all values of x (integers and fractions) that are less than 3.

Example 2

Find the solution set of all numbers for $3 - y < 15$.

Worked solution

$-y < -3 + 15$ (move 3 to the right side of the inequality)
i.e. $-y < +12$

Now multiply both sides by −1 to make y positive, giving
        $y > -12$
Thus the *solution set* is all values of y that are greater than −12.

Exercise 34

1. Find the solution set of numbers for the following inequalities:

(a) $x + 15 > 18$ (b) $16x \geq 80$ (c) $5x - 4 < 16$

(d) $\frac{x}{3} \leq 2$ (e) $2p + 9 > 25$ (f) $\frac{p}{3} + 9 \leq 12$

(g) $2x + 15 > 23$ (h) $-4x + 5 < 25$ (i) $-\frac{x}{4} < 3$

(j) $3y + 2 \geq 17$ (k) $14y - 3 \leq 81$ (l) $-\frac{q}{4} + 11 > 13$

(m) $-3 + \frac{y}{2} < 1$ (n) $5x + 2 > 12 + 3x$ (o) $5 \geq 2x - 3$

(p) $4 - 3p < 12 + p$ (q) $2x + 8 \geq 5x - 4$ (r) $9 - 4y < 6 - 3y$

2. Identify the set of numbers that satisfies the following inequalities:

(a) $x - 10 \leq 30$, where x is a squared number.

(b) $15x < 105$, where x is a positive odd number.

(c) $\frac{p}{4} + 2 < 10$, where p is a positive integer divisible by 6.

(d) $3x - 16 \leq 29$, where $x > 12$ and an integer.

(e) $\frac{m}{5} - 3 \leq 7$, where m is a squared number that is odd.

(f) $5y - 12 \leq 28$, where $y \geq 5$ and an integer.

(g) $-3x - 4 \leq 8$, where $x \leq 0$ and an integer.

(h) $-9 - 5y > 6$, where $y > -7$ and an integer.

(i) $-\frac{1}{4}p + 12 \leq 16$, where $p \leq -10$ and an even integer.

(j) $6 + \frac{1}{3}y \leq 10 + y$, where $y \leq -2$ and an even integer.

3. Graphs of inequalities

The graph of an inequality results in a *solution space* (an area) in the Cartesian plane, where all coordinates within this space will satisfy the inequality relationship.

Method

1. Write the inequality in standard form,
   e.g. $y < \frac{4}{5}x + 2$.
2. Sketch the inequality as an *equation* (i.e. ignore the inequality sign and treat it as an equal sign).
3. Now consider the inequality sign and find the area (above or below, left or right of the straight line equation) that satisfies the inequality condition.

   Hint: Test a coordinate off the line by substituting it into the inequality. If this coordinate satisfies the inequality, then all coordinates in this same area will also satisfy the inequality. This is then the solution space. A useful coordinate to test is the origin (0, 0).

4. If two or more inequalities are plotted, the solution space is the common area where all the separate solution spaces overlap (i.e. this corresponds to solving simultaneous inequalities).

Note:

- If the inequality sign is either < or >, then no point of the solution set lies on the straight line (in this case, sketch the straight line graph as a broken line).

- If the inequality sign is either ≤ or ≥, then points of the solution set also lie on the straight line (in this case, sketch the straight line graph as a solid line).

Example 1

Sketch the graph of $x < 2$.

Worked solution

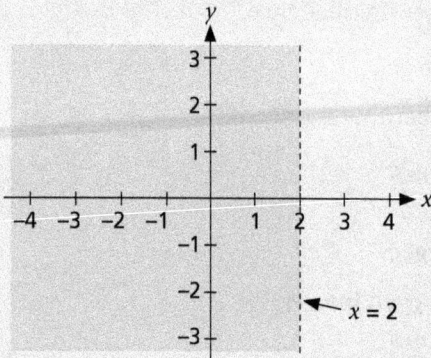

Example 2

Sketch the graph of $x < 1$ and $y > -2$.

Worked solution

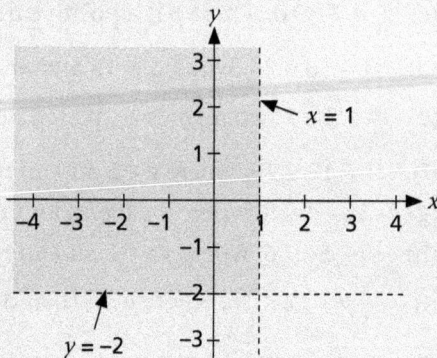

Example 3

Sketch the graph of $y - 2x \geq 1$

Worked solution

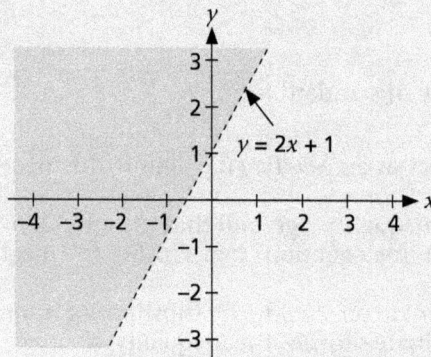

Example 4

Sketch the solution space that satisfies the following inequalities simultaneously:
$$2x + 3y \geq 12 \text{ and } 5x - 2y \leq 8$$

Worked solution

Write each inequality in the standard form:

$2x + 3y \geq 12$ becomes $y \geq -\frac{2}{3}x + 4$ (Inequality [1])

and $5x - 2y \leq 8$ becomes $y \geq \frac{5}{2}x - 4$ (Inequality [2])

Plot each inequality (as if it were an equation).

Identify the solution space for each
inequality separately.
Substitute the origin coordinate (0, 0).

For [1]: Is $2(0) + 3(0) \geq 12$,
i.e. is $0 \geq 12$? The answer is no.
Since (0, 0) lies to the left of the
graph, this coordinate, and all other
coordinates to the left of (below) the
graph will not satisfy this inequality.
Hence the valid solution space of
inequality [1] lies to the right of
(above) the graph.

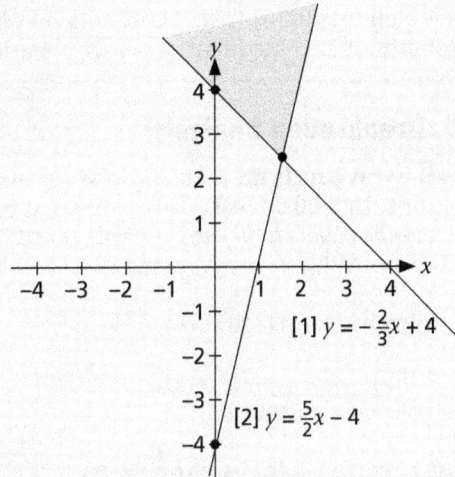
$[1] \ y = -\frac{2}{3}x + 4$
$[2] \ y = \frac{5}{2}x - 4$

For [2]: Is $5(0) - 2(0) \leq 8$,
i.e. is $0 \leq 8$? The answer is yes.
Since (0, 0) lies above (to the left of)
the graph, this coordinate, and all other coordinates above the graph will also
satisfy this inequality.
Hence the valid solution space of inequality [2] lies to the left of the graph.

The *solution space* that satisfies both *inequalities simultaneously* is the common
area (i.e. the overlapping area) to the right (above) of inequality [1] and to the
left of inequality [2] as shown in the sketch.

Exercise 35

1. On a sketch graph, shade the region for the set of points that satisfy the conditions:

    (a) $x < 2$ and $y \leq 4$
    (b) $x \leq 3$ and $y < 1$
    (c) $x < -2$ and $y < 2$
    (d) $x < -1$ and $y \geq 3$
    (e) $x \geq 4$ and $y < -2$
    (f) $x > 1$ and $y \geq 3$
    (g) $x + y < 4$ and $x > 0$, $y > 0$
    (h) $x + y \leq 6$ and $x > 0$, $y > 0$
    (i) $x - y \leq 3$
    (j) $-y + 2x \leq -1$
    (k) $2x + 2y \leq 6$
    (l) $x - 2y \geq 4$
    (m) $2x - 3y \leq 12$
    (n) $-3y \geq 12x - 6$

2. For each pair of inequalities, sketch the graphs and identify if the coordinate (0, 0) is in the solution space:

(a) $5x + y < 28$
     $x - y > 2$

(b) $x + 3y \geq 12$
     $5x - 2y \leq 10$

(c) $4x + 3y \leq 9$
     $5x - 4y > 8$

(d) $-x - y \leq 4$
     $2x + y \geq -2$

(e) $2x + y > 4$
     $3x - y > 4$

(f) $2x + 3y \leq 12$
     $-4y \leq 8x - 4$

(g) $-3x + 2y \geq 4$
     $-2x - 5y \leq -5$

(h) $-2x + y > 4$
     $2x + y > -3$

3. Which quadrant, if any, contains no point (x, y) that satisfies the inequality $2x - 3y \leq -6$?

4. Which quadrant, if any, contains no coordinate (x, y) that satisfies the simultaneous inequalities $y > 2x + 4$ and $-y > 2x + 3$?

2.11 Break-even analysis

Break-even analysis is an application of straight line graphs and simultaneous equations. In a business context it is used to *find* the *threshold sales volume* such that *total revenue equals total costs*. At this point, a company is neither making a loss nor making a profit, but breaking even!

The following terms are used:

Unit selling price = p (in rand)
Sales volume = x (in units)

Total revenue = R (in rand) = px

Total fixed costs = f (in rand) (rental, machinery, admin costs)
Variable cost per unit = v (in rand) (labour, raw materials, electricity)

Total cost = fixed costs + variable costs = $f + vx$

Break-even sales volume (x) is found as follows:

Total revenue (R) = Total costs (fixed + variable)

Find x such that	$px = f + vx$

Solve algebraically for x	$x = \dfrac{f}{(p - v)}$

Both the total revenue equation ($y = px$) and total cost equation ($y = f + vc$) are straight line graphs.

The point of intersection between the total revenue line and the total cost line is the break-even volume.

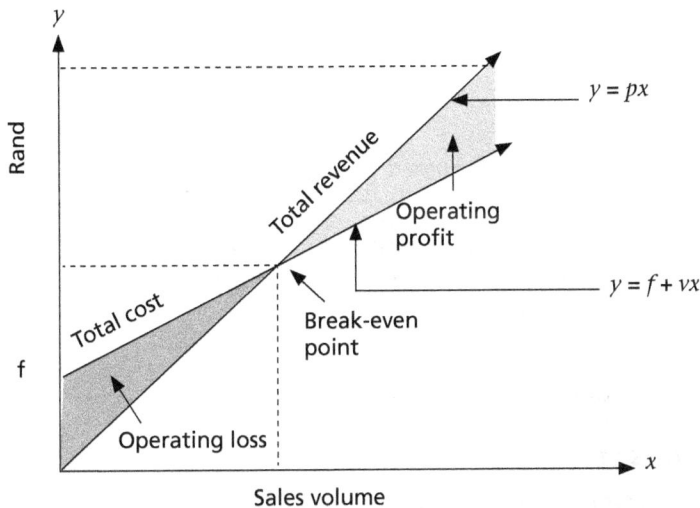

Rand

f

Total revenue

Total cost

Operating profit

Operating loss

Break-even point

$y = px$

$y = f + vx$

Sales volume

x

y

Example

A company sells a product for R8. Its variable unit cost is R5 and total fixed costs are R840.
How many products must the company sell to break even?

Worked solution

Let the break-even quantity = x units
Total revenue is the straight line equation $\quad y = 8x$
Total costs is the straight line equation $\quad\quad y = 5x + 840$
The **break-even quantity** is found when $\quad 8x = 5x + 840$
$$\therefore 3x = 840$$
$$\therefore x = 280 \text{ units}$$

Exercise 36

1. What is a company's break-even sales quantity if the unit selling price is R14, the variable unit cost is R9 and the total fixed cost is R5 000?

2. A company's fixed cost is R3 200 per month. It sells its product for R20 and the variable unit cost is 75% of its unit selling price. How many units must the company sell per month to break even?

3. A company has monthly fixed costs of R5 600. The variable cost of producing one unit is R16. If the company sells its product with a mark-up of 50% on its variable cost per unit, how many units must the company sell per month to break even?

4. Construct the 'total revenue' and 'total cost' straight line graphs for question 1. Sketch these two graphs on a set of axes and show the break-even sales quantity coordinate.

CHAPTER 3

GEOMETRY

In geometry we visualise and analyses spatial problems (areas, volumes, perimeters) using three basic shapes (triangles, rectangles and circles).

3.1 Triangles

A triangle is an area bounded by three sides.

1. Basic properties of a triangle

(a) The *sum* of the angles of a triangle equals 180°.
$\hat{a} + \hat{b} + \hat{c} = 180°$

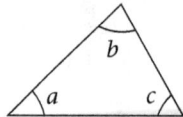

(b) The *largest angle* is always *opposite* the *longest side* (and vice versa).

(c) An *external angle* equals the *sum* of the *two interior opposite* angles.
$\hat{x} = \hat{a} + \hat{b}$.

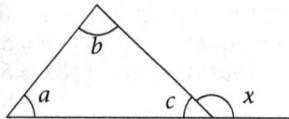

(d) The *longest side* of a triangle can *never equal nor be greater than* the *sum* of the *other two sides*.

(e) The *shortest side* of a triangle can *never equal nor be less than* the *difference of the other two sides*.

2. Types of triangles

Equilateral

Three sides are equal.
All angles = 60°

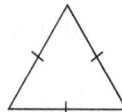

Isosceles

Two sides are equal.
Angles opposite the equal sides are equal.

Scalene

None of the sides are equal.
None of the angles are equal.

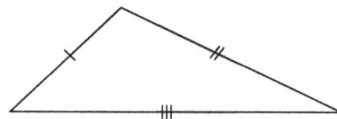

Right-angled triangle

One angle is equal to 90°.

Obtuse-angled triangle

One angle is greater than 90° but less than 180°.

Acute-angled triangle

All three angles are less than 90°.

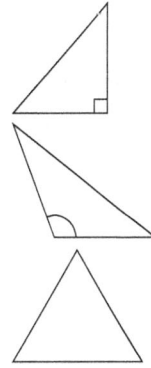

Exercise 37

1. Match the correct name that describes each type of triangle.

   A   Acute-angled and scalene
   B   Obtuse-angled and scalene
   C   Acute-angled and isosceles
   D   Obtuse-angled and isosceles
   E   Right-angled
   F   Right-angled and isosceles
   G   Equilateral

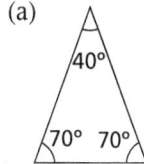

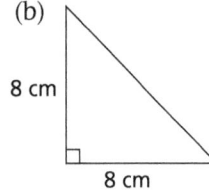

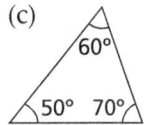

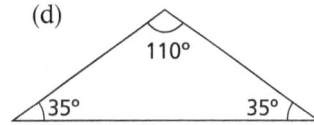

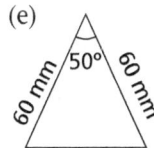

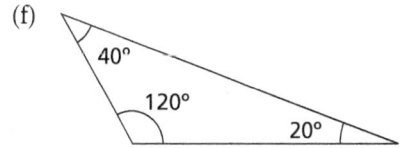

   (a) 40°, 70°, 70°

   (b) 8 cm, 8 cm

   (c) 60°, 50°, 70°

   (d) 110°, 35°, 35°

   (e) 60 mm, 50°, 60 mm

   (f) 40°, 120°, 20°

2. Identify the type of triangle in each case and find the missing angle(s).

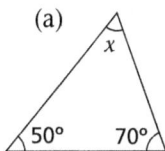

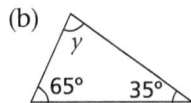

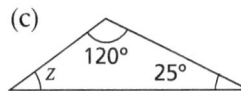

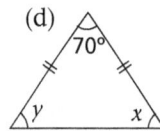

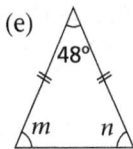

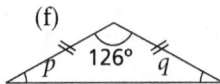

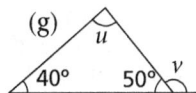

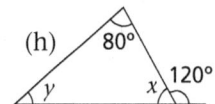

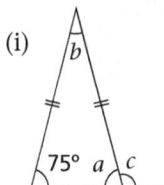

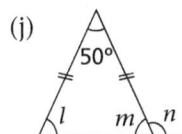

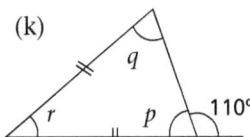

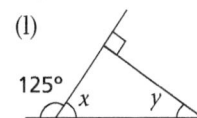

   (a) x, 50°, 70°

   (b) y, 65°, 35°

   (c) 120°, z, 25°

   (d) 70°, y, x

   (e) 48°, m, n

   (f) p, 126°, q

   (g) u, 40°, 50°, v

   (h) 80°, y, x, 120°

   (i) b, 75°, a, c

   (j) 50°, l, m, n

   (k) q, r, p, 110°

   (l) 125°, x, y

3. If 3 and 8 are the lengths of two sides of a triangle, which of the following can be the length of the third side?

   (a) 5   (b) 8   (c) 11

4. Find x:

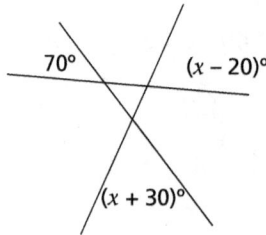

5. In the figure below, PS = QS and angle QPR = 30°. What is the ratio of angle PRS to angle PQR?

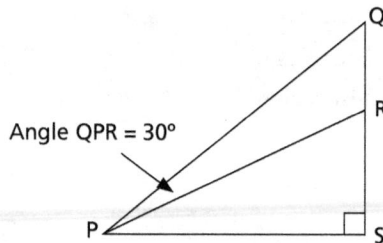

3. Pythagoras' theorem

For any *right-angled* triangle:

(hypotenuse)^2 = (opposite)^2 + (adjacent)^2

$a^2 = b^2 + c^2$

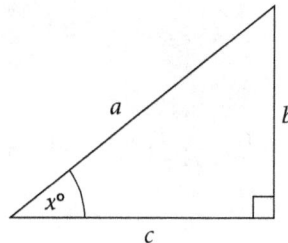

4. Special right-angled triangles

There are a few right-angled triangles that have special relationships (ratios) between their sides. These are worth remembering, as they occur often in GMAT questions. If any of these ratios can be identified (usually as a multiple of the basic ratio), then all sides can immediately be determined.

Note: In the following relationships, the largest ratio is always associated with the hypotenuse. The first three cases are the most common.

Case 1   Ratio of sides $\mathbf{1 : 1 : \sqrt{2}}$

Angles opposite the sides that are equal to 1 are 45° (right-angled isosceles).

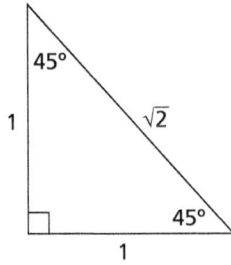

Example:

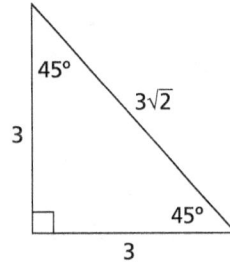

Case 2   Ratio of sides $\mathbf{3 : 4 : 5}$

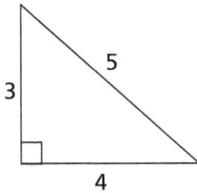

Example:

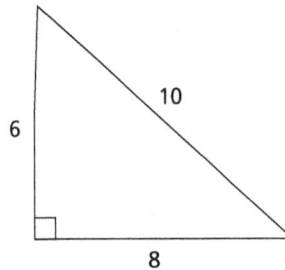

Case 3   Ratio of sides $\mathbf{1 : 2 : \sqrt{3}}$

The angle opposite the side that equals 1 is 30° and opposite the side that equals $\sqrt{3}$ is 60°.

This is a 30°/60° right-angled triangle.

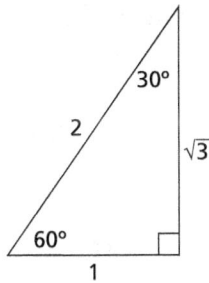

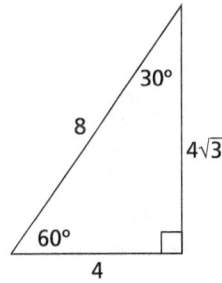

Case 4   Ratio of sides $\mathbf{5 : 12 : 13}$ (this is less common).

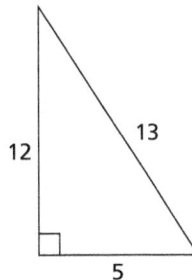

Case 5   Ratio of sides **7 : 24 : 25** (this is also less common).

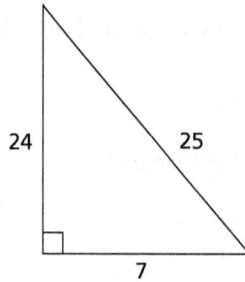

Exercise 38

1. Find the length of the missing side of each of the following *right-angled triangles* (notation: a = hypotenuse; b = opposite; c = adjacent sides).

    Reduce the answer to its simplest form, but leave it in square root form if it is not easy to simplify using mental arithmetic, e.g. if the answer is $\sqrt{50}$ then write it as $5\sqrt{2}$, which is derived from $\sqrt{25 \times 2}$.

    (a) $a = ?$ $b = 6$ $c = 8$ (b) $a = ?$ $b = 5$ $c = 7$

    (c) $a = 5$ $b = 2$ $c = ?$ (d) $a = 6$ $b = 3$ $c = ?$

    (e) $a = 8$ $b = ?$ $c = 5$ (f) $a = 6$ $b = ?$ $c = 4$

    (g) $a = 13$ $b = 5$ $c = ?$ (h) $a = 10$ $b = 8$ $c = ?$

    (i) $a = ?$ $b = 7$ $c = 7$ (j) $a = 18$ $b = ?$ $c = 9$

    (k) $a = ?$ $b = 4$ $c = 4$ (l) $a = 40$ $b = ?$ $c = 32$

    (m) $a = ?$ $b = 12$ $c = 5$ (n) $a = 8$ $b = 4\sqrt{3}$ $c = ?$

    (o) $a = 6\sqrt{2}$ $b = ?$ $c = 6$ (p) $a = ?$ $b = 7$ $c = 7$

2. Find the length of the missing side in the following figures:

(a)

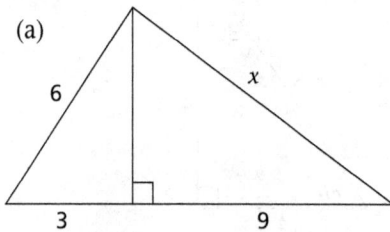

(b)

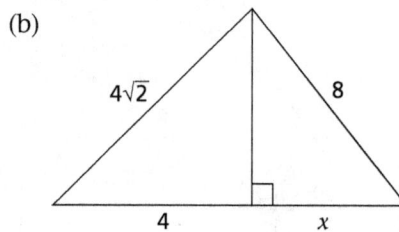

(c)

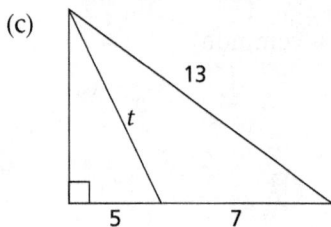

(d)

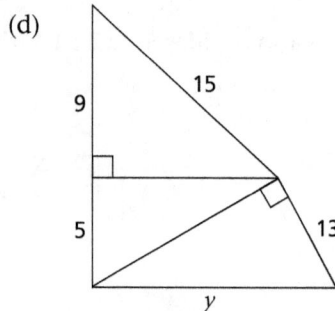

3. For each question, draw a neat sketch and find the length of the missing side:

   (a) An aircraft flies 400 km due West and then a further 300 km due South. How far is the aircraft from its starting point?

   (b) A ladder of length 5 m leans against a vertical wall with its feet 2 m from the base of the wall. How high does the ladder reach up the wall?

   (c) A ship sails 100 km due North and a further x km due East. The ship is then 260 km from its starting point. How far due East did the ship sail?

   (d) A rectangle measures 8 m by 5 m. What is the length of the diagonal?

   (e) A rectangle of length 10 cm has diagonals of length 12 cm. What is the width of the rectangle?

   (f) Draw a diagram of a jewellery box measuring 6 cm (w) by 8 cm (l) by 10 cm (h). What is the length of the longest diagonal of the jewelry box?

   Hint: Confirm the following formula: longest diagonal of a box

   $$= \sqrt{(w^2 + l^2 + h^2)}, \text{ where } w = \text{width}, l = \text{length and } h = \text{height}.$$

3.2 Area of rectangles and triangles

1. Area of a rectangle

Area of a **rectangle** = length × breadth   (A = $l \times b$)

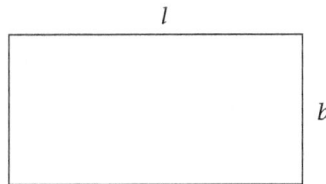

A square is a special case of a rectangle – all its *sides* are *equal*.

Area of a **square** = length^2 (Area = l^2)

2. Area of a triangle

Area of a **triangle** = $\frac{1}{2}$ base × perpendicular height   (A = $\frac{1}{2}bh$)

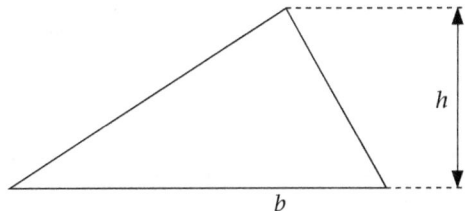

Length of sides and areas of triangle – a useful property

If the lengths of the sides of a triangle are *all changed* by a *constant multiple*, n (×2, ×3, ×4, ×5, etc.), then the *area* of the enlarged triangle is n^2 *bigger* than the original triangle's area.

$$\boxed{\text{Revised triangle area} = n^2 \times \text{original triangle area}}$$

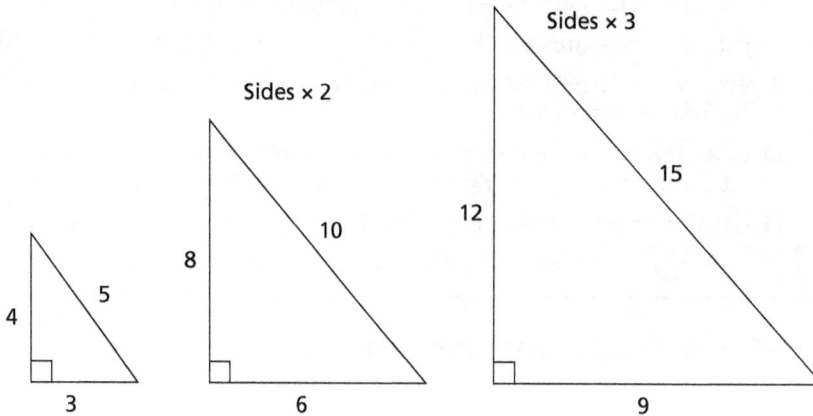

Sides × 3

Sides × 2

15

12

10

8

5

4

3

6

9

Area = $\frac{1}{2}$(3)(4) = 6   Area = $\frac{1}{2}$(6)(8) = 24   Area = $\frac{1}{2}$(9)(12) = 54

Apply formula $\boxed{2^2 \times 6 = 24}$ $\boxed{3^2 \times 6 = 54}$

Note: The same formula applies if the length of the sides of a triangle are *reduced* by a constant multiple n, $\left(n = \frac{1}{2}, \frac{1}{3}, \frac{1}{5} \text{ etc.}\right)$

e.g. if halved $\left(n = \frac{1}{2}\right)$, then the reduced area $= \left(\frac{1}{2}\right)^2 \times$ original area;

if reduced to $\frac{1}{5}$ $\left(n = \frac{1}{5}\right)$, then the reduced area $= \left(\frac{1}{5}\right)^2 \times$ original area.

A useful property of isosceles triangles

A perpendicular line onto the base of an isosceles triangle *bisects* the base (i.e. divides it into two equal halves) and creates two equal-area right-angled triangles.

3.3   Perimeter of rectangles, squares and triangles

Perimeter of *rectangle* = 2 × (length + breadth)	P(rectangle) = 2(l + b)
Perimeter of *square* = 4 × length	P(square) = 4l
Perimeter of *triangle* = sum of the three sides	P(triangle) = $s_1 + s_2 + s_3$

Exercise 39

1. Find the following areas:

    (a) a rectangle with sides of 60 cm and 40 cm.

    (b) a square of length 12 m.

    (c) a rectangle of width 10 cm and length 3 times its width.

    (d) a rectangle of length 30 cm and diagonal 50 cm.

    (e) a square with diagonal $8\sqrt{2}$ m

    Hint: Consider the special right-angled triangles.

    (f) a triangle with base of 7 m and vertical height of 4 m.

    (g) a triangle of perpendicular height 10 cm and base 15 cm.

    (h) a right-angled triangle with the opposite side 15 m and the hypotenuse 25 m.

    (i) a triangle with base split into 5 m and 3 m segments by a perpendicular onto the base, and a side of 5 m that is adjacent to the 3 m line segment of the base.

2. Find the areas of the following shapes:

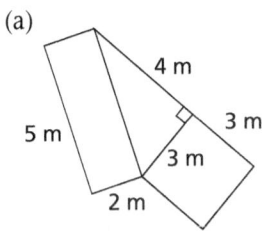

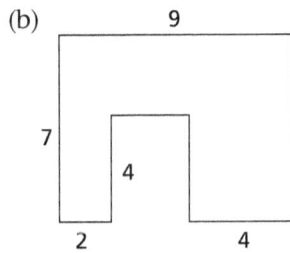

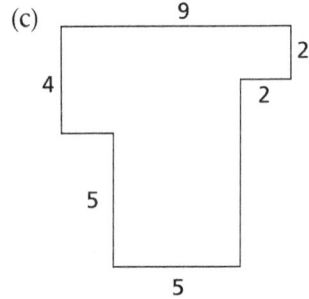

    (a)   (b)   (c)

3. What is the area of a triangle with coordinates:

    (a) $A(7, 4)$, $B(0, 3)$ and $C(4, 0)$ (b) $X(4, 0)$, $Y(6, 3)$ and $Z(6, -3)$

4. The perimeter of a right-angled isosceles triangle is $16 + 8\sqrt{2}$. What is the length of the longest side?

5. Find the area of a right-angled triangle with a perimeter equal to $12 + 4\sqrt{3}$ and one angle equal to 30°.

6. The size of a flat-screen television is given as the length of the screen's diagonal. How much larger is the area of a square 21-inch flat-screen compared to a square 19-inch flat-screen?

7. Find the perimeter of the following shapes:

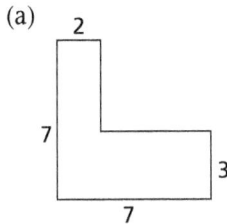

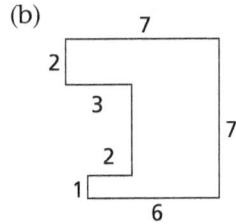

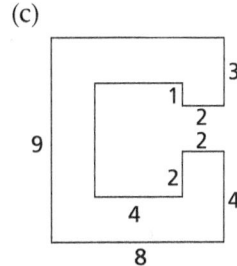

    (a)   (b)   (c)

8. A flower bed is in the shape of a right-angled triangle ABC with angle B = 90°. AC = z metres, AB = x metres and BC = y metres. If the area of the flower bed is 24 square metres, and $x = (y + 2)$, what is the value of z?

3.4   Circles

A circle has a 360° angle at its center (a full revolution).

1. Area of a circle

> The area of a circle, A = πr^2
>
> where $\pi = \frac{22}{7}$ and r = radius

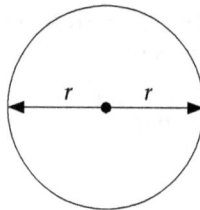

Example
Find the area of a circle with a radius of 7 m.

Worked solution
Area = $\pi(7)^2 = \pi(49) = 49\pi$ m^2

2. Area of a sector of a circle

The area of a sector of a circle in relation to the circle's area, is *directly proportional* to the size of the angle x at the centre of the circle that supports it.

> Area of a sector = $\frac{x}{360} \times \pi r^2$

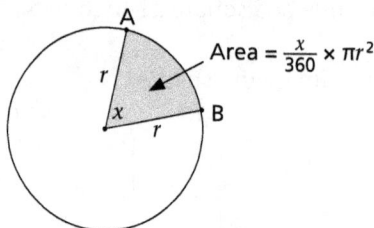

Area = $\frac{x}{360} \times \pi r^2$

Example 1
What is the *area* of a sector of a circle with an angle of 60° at the centre and a radius of 12 cm?

Worked solution

Area of circle = $12^2\pi = 144\pi$.

Proportion of 60° (angle of sector) to a full revolution of 360° = $\frac{60}{360} = \frac{1}{6}$

Area of 'slice' = $\frac{1}{6}$ of $144\pi = 24\pi$

Example 2

What is the *angle* at the centre of a circle that supports a sector with an area of 24π (cm^2) and a radius of 12 cm?

Worked solution

Area of circle = $12^2\pi = 144\pi$.

Proportion of 24π (area of sector) to the area of the circle = $\frac{24\pi}{144\pi} = \frac{1}{6}$

Angle at centre = $\frac{1}{6}$ of 360° = 60°

3. Circumference (perimeter) of circle

> The **circumference** of a circle, $C = 2\pi r$ or πd
> where d = diameter (= 2 × radius).

Example

Find the circumference of a circle with radius of 14 cm.

Worked solution

Circumference = $2\pi(14) = \pi \times 28 = 28\pi$ cm

4. Length of an arc of a circumference

An arc is a segment of the circumference. In the circle diagram on the previous page, the line segment AB on the circumference of the circle is an **arc**.
Its length is proportional to the size of the angle that supports it.

$$l \text{ (arc)} = \frac{x}{360} \times C$$

Example 1

If the sector of a circle with radius 12 cm has an angle of 60° at the centre, what is the length of the arc that supports it?

Worked solution

If the angle at the centre is 60°, then the length of the arc is $\frac{1}{6}$ of its circumference.

Circumference = $(2\pi \times 12)$ cm = 24π

Therefore the length of the arc is $\frac{1}{6} \times (24\pi) = 4\pi$ cm

Example 2

If the sector of a circle with radius 30 m has an angle of 90° at the centre, what is the length of the arc that supports it?

Worked solution

If the angle at the centre is 90°, then the length of the arc is $\frac{1}{4}$ of the circumference.
Circumference = $(2\pi \times 30)$ m = 60π
Therefore the length of the arc is $\frac{1}{4} \times (60\pi) = 15\pi$ m

5. Properties of a triangle inside a circle

Property 1

A triangle that is *inscribed* in a circle (i.e. one side of the triangle is the diameter of the circle and the opposite angle lies on the circumference) is a *right-angled* triangle.

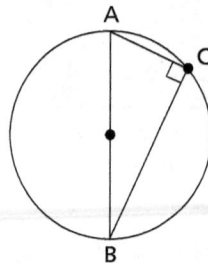

Property 2

In a rectangular coordinate system, a triangle with a right angle (90°) at the origin of a circle (0, 0) and the other two coordinates on the circle circumference will have *transpose circumference coordinates*.

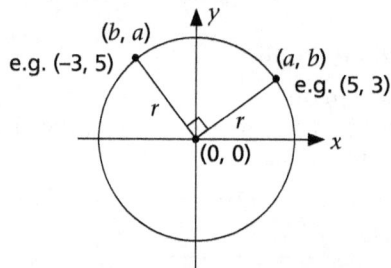

Property 3

The angle at the centre of a circle is *twice* the angle at the circumference when supported by the same vertices on the circumference.

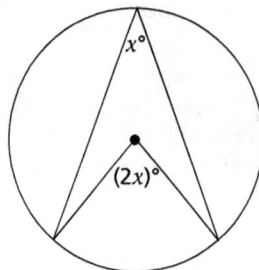

Exercise 40

1. Find the following areas of the circles (leave answers in π form):
   (a) radius = 5 cm
   (b) radius = 7 km
   (c) diameter = 12 m
   (d) radius = 1.5 km
   (e) diameter = 2.6 cm
   (f) radius = $2\frac{1}{2}$ m

2. Find the circumferences of the following circles (leave answers in π form):
   (a) radius = 5 cm
   (b) radius = 7 km
   (c) diameter = 42 m
   (d) radius = 1.5 km
   (e) diameter = $2\frac{6}{11}$ cm
   (f) radius = $3\frac{1}{2}$ m

3. What is the length of the arcs for the given angles for the following circles?
   (a) angle of arc = 30° and radius = 4 cm
   (b) angle of arc = 45° and radius = 24 m
   (c) angle of arc = 240° and radius = 75 mm

4. What is the area of a sector of a circle with the given radius (or diameter) and angle at the centre?
   (a) angle at centre = 30° and radius = 6 cm
   (b) angle at centre = 45° and radius = 1.2 km
   (c) angle at centre = 120° and diameter = 18 mm

5. What is the angle at the centre of a circle with the given radius (or diameter) and area of a sector?
   (a) area of sector = 27π (m²) and radius = 9 m
   (b) area of sector = 25π (cm²) and diameter = 30 cm
   (c) area of sector = 12π (mm²) and diameter = 8 mm

6. AC and BC are two bridges of equal length over a circular pond O. AB is the diameter of the pond and C is a point on the pond's circumference. Find the ratio of walking the bridges (AC and BC) relative to walking around the edge of the pond from A to B.

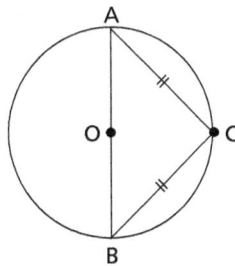

7. A conveyer belt is attached to two circular pulleys at each end. The belt is 15 m in length and each pulley is 1m in diameter. What is the distance between the centres of the two pulleys?

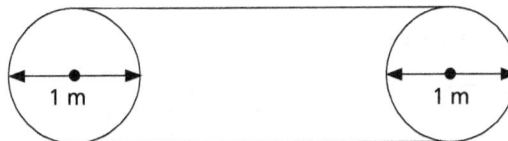

8. An equilateral triangle ABC is inscribed in a circle with its vertices A, B and C on the circumference of the circle. If arc ABC = 24π, what is the area of the circle?

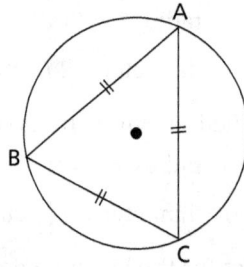

9. A circle O has two coordinates P($-\sqrt{3}$, 1) and Q(s, t) on its circumference. Coordinate Q is in the first quadrant. Angle POQ is a right angle. What is the value of s?

10. Circle O has diameter CD = 18 cm. Line segments AB and CD are parallel. Angle ADC = 35°. Find the length of the arc AB.

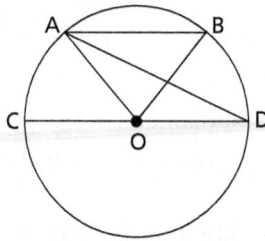

11. A point at the coordinate A (–3, 5) on the circumference of circle O, centred at (0, 0), is rotated clockwise through 270° to coordinate B. What is the coordinate reference at B?

12. A point at the coordinate M (4, –1) on the circumference of circle O, centred at (0, 0), is rotated anti-clockwise through 180° to coordinate N. What is the coordinate reference at N?

13. P (–6, 8) is a coordinate on the circumference of circle O, centred at (0, 0). What is the diameter of circle O?

3.5 Cuboids

A cuboid is a box-shaped container.
We are mainly interested in the volume of this shape.

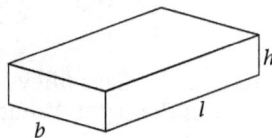

Volume of a cuboid (box shape)

$$V = l \times b \times h$$

(i.e. length × breadth × height)

Note: The volume of a cuboid is expressed in cubed units, e.g. mm^3, cm^3, m^3, km^3.

> **Example**
>
> Find the volume of a box-like container that is 6 m long, 3 m wide and 4 m high.
>
> **Worked solution**
>
> Volume (V) = 6 × 3 × 4
> $\qquad$ = 72 m^3

3.6 Cylinders

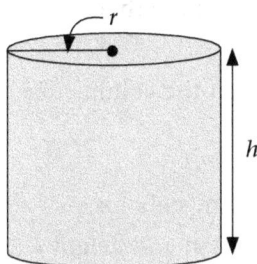

Cylinder

Volume of a cylinder = $\pi r^2 h$, where r = radius and h = height
Surface area of a cylinder (including top and bottom) = $2\pi r^2 + 2\pi rh$
Surface area of a cylinder (only cylindrical side) = $2\pi rh$

Cylinder dimensions and cylinder volumes – a useful property

If both the radius and the height of a cylinder are *increased* by a *constant multiple*, n (×2, ×3, ×4, ×5, etc.), then the *volume* of the enlarged cylinder is n^3 *bigger* than the original cylinder's volume:

e.g. if both radius and height are doubled (×2), volume increases 2^3 (8 times bigger); if trebled (×3), volume increases by 3^3 (27 times bigger).

Note: $\quad$ The same property holds if both the radius and height are *reduced* by a constant multiple, $n \left(n = \frac{1}{2}, \frac{1}{3}, \frac{1}{4}\right)$, etc.:

$\qquad$ e.g. if $n = \frac{1}{2}$, then the reduced volume is $\left(\frac{1}{2}\right)^3$ of its original volume.

The following are general rules: (let n be the scaling factor):

Rule 1: If *both* the *radius* and the *height* are scaled by the same factor, n, then the *new volume* is V$_2$ = n^3 × V$_1$ (i.e. it changes by the *cube* of the scaling factor).

Rule 2: If only the *radius* (and not the height) is scaled by a factor, n, then the *new volume* is $V_2 = n^2 \times V_1$ (i.e. it changes by the *square* of the scaling factor).

Rule 3: If only the *height* (and not the radius) is scaled by a factor, n, then the *new volume* is $V_2 = n \times V_1$ (i.e. it changes only by the scaling factor).

Example

Cylinder A has a radius of 2 m and a height of 3 m. If both the radius and the height were doubled to produce cylinder B, how many times larger (by volume) is cylinder B relative to cylinder A?

Worked solution

Cylinder A: volume $= \pi \times 2^2 \times 3 = 12\pi$ (m³)
Cylinder B: volume $= \pi \times 4^2 \times 6 = 96\pi$ (m³)
Cylinder B is 8 times larger (by volume) than cylinder A $\left(\frac{96}{12} = 8 \text{ times}\right)$

Note: $8 = 2^3$ – the multiple by which the volume of a doubled cylinder increases.

Exercise 41

1. Find the volumes of the following cuboids:
   (a) length = 3 cm, breadth = 6 cm and height = 2 cm.
   (b) length = 5 m, breadth = 5 m and height = 5 m.
   (c) length = $\frac{3}{8}$ m, breadth = $2\frac{2}{3}$ m and height = $\frac{1}{2}$ m.
   (d) If the volume of a box is 42 cm³, what is its length, if the breadth is 4 cm and the height is 2 cm?
   (e) If the volume of a rectangular tank (cuboid) is 180 m³, how high is the tank if its length is 30 m and its breadth is 12 m?
   (f) How much water is required to fill a pool that is 8 m long, 3.5 m wide and 1.5 m deep when it requires 1 000 ℓ of water to fill one cubic metre?

2. One cylinder has a capacity of 100 ℓ. A second cylinder has twice the radius and twice the height of the first cylinder. If the contents of the first cylinder is poured into the second cylinder, how much unused capacity remains in the second cylinder?

3. How many cylindrical cans with diameter 6 cm and height 8 cm would be needed to hold the contents of a full cylindrical can with radius 9 cm and height 24 cm?

4. A cylindrical concrete dam is to be waterproofed (sides and base only). The dam has a diameter of 8 m and a height of 3 m. If 1 ℓ of waterproofing paint covers 2π m² of surface area, how many litres of waterproofing paint must be bought?

5. One cylinder has a radius of 10 cm and is full of liquid. A second cylinder has half the radius of the first cylinder and the same height. What fraction of the first cylinder's contents can be poured into the second cylinder?

6. An 8 m length of piping with a constant diameter is cut into two lengths of 2 m and 6 m. Each length of piping is sealed at one end. If 180 cl of a liquid is used to fill the 6 m long sealed pipe, how much of the same liquid is needed to fill the 2 m long sealed pipe?

4

BASIC STATISTICS

Statistics is a discipline of techniques to summarise sample data and compute probabilities to aid management decision making.

This chapter looks at probabilities and a few basic statistical tools (called descriptive statistical techniques) that can be used to describe data.

Sample data can be described by:

- a *central location* measure using either an *average* (arithmetic mean) or a *median*; and
- a measure of *spread* (or dispersion) about the central location using a *standard deviation*.

4.1 Central location measures (average, weighted average, median)

1. Arithmetic mean (average) ($\bar{x}$)

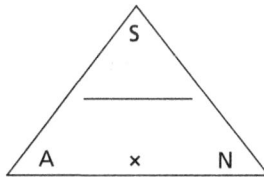

S = sum of all the numbers
A = average ($\bar{x}$)
N = number of values counted

Mean (average) is a *central number* for a set of numbers.

$$\text{Arithmetic mean} = \frac{\text{sum of all the numbers}}{\text{number of values counted}}$$

$$\bar{x} = \frac{\Sigma x}{n}$$

where Σx is the sum of all the numbers and n is the number of data values counted.

Example

Find the average age of the following five boys: 9, 14, 10, 8 and 14 years.

Worked solution

Sum of the five ages:
$\Sigma x = 9 + 14 + 10 + 8 + 14 = 55$ years (total)
To find the average, divide the total by $n = 5$
Average age $= \frac{55}{5} = 11$ years

Note: To revise averages, always compute the *sum of all the numbers* first.

> **Example**
>
> The average age of four boys is 12.4 years. If another boy joins the group, the new average of the group of five boys is 13 years.
> How old is the boy?
>
> **Worked solution**
>
> Find the *sum of the ages* of the *four* boys:
> Sum of the ages = 4 × 12.4 = 49.6 years
> Next, find the *sum of the ages* of the *five* boys:
> Sum of the ages = 5 × 13 = 65 years
> The *difference* in the *sum of the ages* is the age of the additional (5th) boy:
> 5th boy's age   = (65 − 49.6) years = 15.4 years

Exercise 42

1. Find the arithmetic average of the following:

    (a) The lengths of holidays of 6 families: 17 days, 22 days, 10 days, 15 days, 30 days, 14 days.

    (b) The time to deliver 4 parcels: 35 min, 40 min, 28 min and 41 minutes.

    (c) The price of 7 items at a store: R3.50, R6.25, R8.75, R2.50, R5.25, R8.50 and R7.25.

    (d) The weight of 5 tins of food: 1.2 kg, 2.4 kg, 0.5 kg, 1.8 kg and 1.1 kg.

    (e) The distance (in km) travelled by 6 cars on a litre of fuel: 12.5, 8.6, 10.8, 11.7, 10.9 and 11.5.

2. Solve:

    (a) The average of six numbers is 8.5. When one number is discarded, the average of the remaining numbers becomes 7.2. What is the discarded number?

    (b) The arithmetic mean of five numbers is 11. When one number is discarded, the average of the remaining numbers is 12. What is the discarded number?

    (c) The average of 10, 30 and 50 is 5 more than the average of 20, 40 and x. Find x.

    (d) If x magazines cost R9 each and y magazines cost R15 each, what is the average cost per magazine?

    (e) If $x + y = 8z$, find the average (arithmetic mean) of x, y and z in terms of z.

    (f) If x books cost R5 and y books cost R8, find the average cost per book.

3. Find the expression for the following problem:

    If a basketball team scores an average of x points per game for n games and scores y points in its next game, what is the team's average score for the $n + 1$ games in terms of x and y?

4. The average of seven numbers is 12.2. If the sum of four of these numbers is 42.8, what is the average of the other three numbers?

5. If the average of x and y is 60, and the average of y and z is 80, what is the value of $z - x$?

6. The average male experience in a company is 9.8 years. The average female experience is 9.1 years. If there are 52 males in the company and the combined average experience is 9.3 years, what is the ratio of the number of males to the number of females?

7. Thabo wrote five exams. His scores on the first four exams (in order of completion) were 76, 87, 73 and 96. His average score on the last three exams was 89. What was his average score for the five exams?

8. What is the average (mean) of seven consecutive integers where the average of the first five integers is 8?

9. For the past n days, the average daily production at a factory was 50 units per day. If today's production of 90 units raises the average to 55 units per day, what is n?

2. Weighted average

A weighted average gives more importance (weight) to certain values in an average calculation.

Example

If 10 workers each earn R40 per hour and 15 workers each earn R50 per hour, find the average earnings per hour for all the workers.

Worked solution

S = total earnings for both sets of workers
  = 10 × R40 + 15 × R50 = R1 150

N = total number of workers
  = 10 + 15 = 25 workers

A = Average = $\frac{S}{N}$
  = $\frac{1150}{25}$ = R46

Note 1   In the average calculation, the weight of the 10 workers is $\frac{10}{25}$ and the weight of the 15 workers is $\frac{15}{25}$. The average can also be found by weighting each hourly rate by their respective fractions:
i.e. average = R40 × $\left(\frac{10}{25}\right)$ + R50 × $\left(\frac{15}{25}\right)$ = R46.

Note 2   The average earnings per hour are weighted towards the larger group of workers.
As a *general rule*, a weighted average is always *weighted towards* to the value of the measure linked to the *larger sample size*.

Note 3   If the two sample sizes are equal, then the weighted average will equal the simple average of the two measures.

3. Median

The median represents the *middle number* in a set of *ordered* (ranked) numbers.

Method

1. Sort the data in either ascending or descending order.
2. Divide the sample size, n, by 2 to find the position of the middle number.

   If n is odd, use $\frac{n+1}{2}$. If n is even, use $\frac{n}{2}$.
3. The median is the number in the $\left(\frac{(n+1)}{2}\right)^{\text{th}}$ position, if n is odd.

   The median is the average of the two middle numbers, if n is even.

> **Example**
>
> Find the median age of seven employees: 45, 28, 56, 32, 48, 42, 30.
>
> **Worked solution**
>
> Rank the employee ages: 28, 30, 32, 42, 45, 48, 56
>
> Since $n = 7$ is odd, the middle position is the $\frac{7+1}{2} = 4^{\text{th}}$ position
>
> The employee in the 4th (ranked) position has an age = 42 years (*median* age)

The **median** is also called the **50th percentile**, since the lower 50% of the numbers lies below the median and the upper 50% of the numbers lie above the median.

Exercise 43

1. Use a weighted average approach to solve each of the following:

   (a) In a shipment of 120 parts, 5% were defective. In a shipment of 80 parts, 10% were defective. What is the average defective rate for the two shipments combined?

   (b) 300 seeds were planted in one plot and 200 seeds were planted in a second plot. If 25% of the seeds in the first plot germinated and 35% of the seeds in the second plot germinated, what is the average germination rate across both plots?

   (c) 81% of a sample of 100 drivers in city 1 wore seat belts and only 60% of a sample of 200 drivers in city 2 wore seat belts. What is the average percentage of drivers surveyed from both cities together who wear seat belts?

2. Solve:

   (a) Find the median travelling time (in min) to work for six trips: 49, 40, 46, 48, 40 and 42.

   (b) Five tins of peach jam respectively weigh: 254, 255, 246, 252 and 248 g. What is the median weight (in g) of the five tins?

   (c) The scores in a history test are: 40, 45, 45, 50 , 50, 60, 70, 75, 95, 100. The total score is 630. How many learners achieved more than the median score but less than the mean (average) score?

   (d) If the median of five distinct numbers (14, 9, 12, 10 and x) is 10, which *one* of the following statements is true?

   (i) $x < 10$ (ii) $x \leq 10$ (iii) $x = 10$ (iv) $x > 10$ (v) $x \geq 10$

(e) If $0 < x < 1$, what is the median of x^0, x^1, x^2, $\sqrt{x}$ and x^3?

(f) If $x > 1$, what is the median of x^0, x^1, x^2, $\sqrt{x}$ and x^3?

(g) A list of ordered numbers is 4, 5, 6, 8, 10 and x. If the median is $\frac{6}{7}$ times that of the arithmetic mean of these numbers, what is x?

(h) What is the average of eleven *consecutive* integers if the median of the first nine integers is 7?

(i) Of the following numbers, which one is third greatest?

   (i) $2\sqrt{2} - 1$ (ii) $\sqrt{2} + 1$ (iii) $1 - \sqrt{2}$ (iv) $\sqrt{2 - 1}$ (v) $\sqrt{2}$

4.2 Standard deviation (S)

Standard deviation is a measure of *average dispersion* (spread) of numbers around the mean of the numbers. Calculation is not required (the formula is given for information only):

$$S = \sqrt{\frac{\Sigma(x - \bar{x})^2}{(n - 1)}}$$

Interpretation (by example):

If a sample of 150 men has a mean age of 37 years with a standard deviation of 3 years, then the following interpretation can be given on the assumption that their ages are *normally* distributed (bell-shaped) about the mean of 37 years:

(a) within *one standard deviation* about the mean: i.e. [37 – 3; 37 + 3], the interval 34 to 40 years *covers approximately 68.3%* of all ages from this group

(b) within *two standard deviations* about the mean: i.e. [37 – 2(3); 37 + 2(3)], the interval 31 to 43 years *covers approximately 95.5%* of all ages from this group

(c) within *three standard deviations* about the mean: i.e. [37 – 3(3); 37 + 3(3)], the interval 28 to 46 years *covers approximately 99.7%* of all ages from this group.

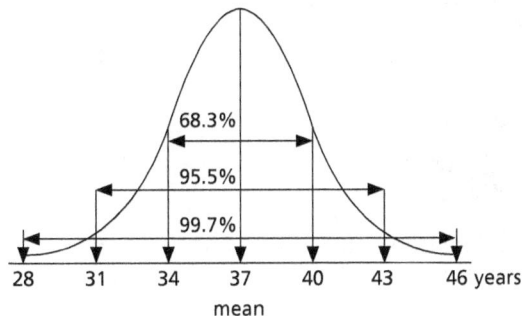

Exercise 44

1. For each of the following sample means and standard deviations, find the one, two and three standard deviation limits about the mean value and interpret them:

    (a) a mean of 24 kg and a standard deviation of 3 kg.

    (b) a mean of 125 km and a standard deviation of 10 km.

    (c) a mean of 72 ℓ and a standard deviation of 2.5 ℓ.

    (d) a mean of 1.05 ℓ and a standard deviation of 0.025 ℓ.

2. The score was 58 when two standard deviations below the mean, and 98 when three standard deviations above the mean. What was the mean score?

3. The mean and standard deviation of a normal distribution is 15 and 3, respectively. What value is exactly 1.5 standard deviations below the mean?

4. If the number of hours per week that residents in a town spend watching TV has a mean of 21 hours and a standard deviation of 4 hours, what is the number of hours per week that a particular resident spent watching TV if she lay $2\frac{1}{2}$ standard deviations below the mean?

5. A list of 100 data values has an average of 6 and a standard deviation of d (d is positive). Which one of the following pairs of data values, when added to the list, *must* result in the list of 102 data values with a standard deviation *less than d*?

    (−6, 0); (0, 0); (0, 6); (0, 12); (6, 6)

4.3 Sets and probability

A **set** represents a grouping of *similar objects*.
The *members of a set* are called **elements** and are written within curly brackets {...}.

A **probability** is the **likelihood of choosing** a **subset** of *elements* from the *set*.

Illustration:

The set of all odd positive numbers less than 10 is written as: P = {1, 3, 5, 7, 9}
The set of all vowels is written as: V = {a, e, i, o, u}
The set of all card suites in a pack of cards is: C = {hearts, clubs, spades, diamonds}
The set of all multiples of 5 less than 40 is written as: M = {5, 10, 15, 20, 25, 30, 35}

1. Terms: events, outcomes and probabilities

An **event** is a set of elements that possesses a particular characteristic of interest (e.g. gender, family size, age of car).

Outcomes are the different members (or elements) of an event (e.g. gender = {male, female}; family size = {0, 1, 2, 3, 4 ... children}; age = {1, 2, 3, 4 ... years old}).

2. Probability

Probability is the likelihood (chance) of a given event occurring.

It is defined as

$$P(A) = \frac{r}{n}$$

where r = number of outcomes of event A
$\quad\quad n$ = total number of all outcomes (called the *sample space*).

Example 1

If there are 15 electrical engineers, 24 civil engineers and 11 mechanical engineers in an electronics company, what is the probability of randomly selecting a civil engineer from the group of all engineers?

Worked solution

Let event A = selecting a civil engineer
r = 24 civil engineers in the company
n = 50 engineers in the company in total
Then P(selecting a civil engineer) = P(A) = $\frac{24}{50}$ = 0.48 (a 48% chance)

Example 2

Find the probability of selecting a flat-dweller from the following sample of 80 residents surveyed in a recent study in Cape Town.

Worked solution

Type of dwelling (events)	Number (outcomes)
House	44
Flat	28
Townhouse	8

Let event A = selecting a flat-dweller from the sample of residents surveyed
r = 28 flat-dwellers
n = 80 residents
Then P(selecting a flat-dweller at random) = P(A) = $\frac{28}{80}$ = 0.35 (a 35% chance)

Exercise 45

1. Find the following probabilities:

   (a) throwing a four on a single roll of a die.

   (b) selecting the winning horse called 'Majesty' in a race of eight horses.

   (c) In an insurance company there were 28 section heads, 14 department heads and 8 division heads. If one of these heads is selected at random to attend a conference, what is the probability that it will be a department head?

2. In a department of 64 persons, 28 staff members have a matric certificate, 24 have a diploma and the remainder have a degree. What is the probability that a randomly selected member of this department will have a degree?

3. In a tour group of 120 tourists, 72 are English-speaking, 24 speak German, 18 speak French and the rest speak Spanish. What is the probability that a randomly selected tour member speaks:

   (a) Spanish   (b) German   (c) German or French?

Note: Sets and probabilities will only be examined further in terms of a *relationship between two events*. Such sets of outcomes are shown in a *cross-tabulation table* (2 × 2 table) for ease of calculation.

3. Cross-tabulation tables

A **cross-tabulation** is a table (usually a 2 × 2 table) that shows how many elements in a sample belong to each outcome of two events.

Example

A sample of 40 people were surveyed and asked where they live (house or flat) and their marital status (married or single). 15 live in a house, 22 are single and 8 are married and live in a house.

(a) How many married persons live in a flat?
(b) What percentage of all persons surveyed does this represent?
(c) What percentage of persons surveyed live in flats?
(d) What is the probability of randomly selecting a person who is both single and living in a house?

Worked solution

Event 1 is *type of accommodation* with outcomes {house, flat}.
Event 2 is *marital status* with outcomes {married, single}.

The numbers for the different outcomes for each event can be shown in a 2 × 2 cross-tabulation table:

		Type of accommodation		
		House	Flat	Total
Marital status	Married	8	?	?
	Single	?	?	22
	Total	15	?	40

The missing values can be found by addition or subtraction:

		Type of accommodation		
		House	Flat	Total
Marital status	Married	8	10	18
	Single	7	15	22
	Total	15	25	40

Once the table is completed, the answers can be read directly from the table.

(a) Married persons living in flats = 10 persons
(b) Percentage of married persons living in flats = $\frac{10}{40} \times 100 = 25\%$
(c) Percentage of persons living in flats = $\frac{25}{40} \times 100 = 62.5\%$
(d) Probability (single and lives in a house) = $\frac{7}{40} \times 100 = 17.5\%$ chance

Note 1: Question (d) is an example of a **joint probability**, where two events occur simultaneously.

Note 2: The keyword **AND** identifies joint probability events.

Finding conditional probabilities – using the cross-tabulation table

When *event A* depends upon the *occurrence* of a prior *event B*, then the probability of event A occurring *given* that event B has already occurred is called a **conditional probability**.

> **Example 1**
>
> If a person is *married*, what is the probability that this person lives in a *house*?
>
> **Worked solution**
>
> There are only 18 married persons.
> Given that a person is *married*, the *probability* that such a person lives in a house $= \frac{8}{18} = 44.4\%$.

> **Example 2**
>
> If a person is a *flat*-dweller, what is the probability that this person is *single*?
>
> **Worked solution**
>
> There are only 25 flat-dwellers.
> Given that a person is a flat-dweller, the probability that such a person is *single* $= \frac{15}{25} = 60\%$.

Addition rule of probability (keyword OR) – using the cross-tabulation table

For two events, when *either one or the other or both* can occur together, the probability of this happening can be found using the **addition rule of probability**.

Stated more formally: the probability of *either event A or event B (or both)* occurring simultaneously is found using the *addition rule of probability*.

Addition rule: $P(A \text{ or } B \text{ (or both)}) = P(A) + P(B) - P(A \text{ and } B)$
(for non-mutually-exclusive events)
If the two events A and B can occur simultaneously, then events A and B are non-mutually-exclusive events.
or
Addition rule: $P(A \text{ or } B) = P(A) + P(B)$
(for mutually exclusive events)
If the two events A and B cannot occur simultaneously, then events A and B are mutually exclusive events.

Hint: Look for the keyword '*or*'. This implies the use of the addition rule of probability.

Example

What is the probability that a surveyed person is *either* married *or* lives in a house (or both)?

Worked solution

P(*either* married *or* lives in a house (or both))
= P(married) + P(lives in a house) − P(married and lives in a house)
= $\frac{18}{40} + \frac{15}{40} - \frac{8}{40} = \frac{25}{40}$ = 62.5% chance

Since a person can be married and live in a house simultaneously, these two events are not mutually exclusive.

Exercise 46

1. Solve:

   (a) In a class of 30 learners, 18 are boys. 20 learners altogether play sport. 7 girls do not play sport.

      (i) How many boys play sport?

      (ii) What is the probability that a randomly selected pupil is a boy who plays sport?

   (b) In a tennis and squash club of 120 members, 72 play squash and 76 play tennis. How many play both games, provided that everyone plays at least one game?

   (c) In a sample of 200 consumers, 150 liked tea, 75 liked coffee and 11 liked neither. What percentage of consumers liked both tea and coffee?

   (d) Of 60 applicants for a job, 28 had at least 4 years experience, 36 had degrees and 6 had less than 4 years of experience and did not have a degree.

      (i) What is the probability that a randomly selected applicant has at least 4 years experience *and* a degree?

      (ii) What is the likelihood that a randomly selected applicant has either less than 4 years experience *or* no degree (*or* both)?

      (iii) Of applicants without degrees, what is the probability that such applicants will have at least 4 years of experience?

   (e) Thirty per cent of the members of a swimming club have passed a lifesaving test. Among the members who had not passed the test, 12 had taken the preparation course and 30 had not taken the preparation course. How many members are there in the swimming club?

   (f) A car hire company has only small- and medium-sized cars in its fleet. Last week it hired out $\frac{3}{4}$ of its cars, including $\frac{2}{3}$ of its small cars. If $\frac{3}{5}$ of its fleet consists of small cars, what percentage of the fleet that were *not hired out* were small cars?

   (g) If 75% of a class answered the first question on a certain test correctly, 55% answered the second question correctly, and 20% answered neither of the questions correctly, what percentage answered both correctly?

(h) Of 60 employees in a company, 36 have a degree, 30 are female and 12 are females with a degree. What is the probability that a randomly selected employee:

   (i) is a female?

   (ii) is a male without a degree?

   (iii) is either a female *or* has no degree?

   (iv) If a male employee is selected, what is the probability that this employee does not have a degree?

(i) In a room, there are 160 people, of whom 15% are women. A group, of whom 30% are women, leaves the room. Of the remaining people in the room, 10% are woman.

   (i) How many persons left the room?

   (ii) Given that a woman is selected, what is the probability that this person stayed in the room?

(j) In a nationwide poll, N people were interviewed. If $\frac{1}{4}$ of them answered 'yes' to question 1, and of those, $\frac{1}{3}$ answered 'yes' to question 2, how many people did not answer 'yes' to both questions? Express the answer in terms of N.

(k) A wine merchant has 144 wine bottles in his store. 90 of them are locally made wines and 48 are red wines. 42 of these bottles are imported white wines. For a randomly chosen bottle of wine, what is the probability that:

   (i) it is a bottle of white wine?

   (ii) it is a locally made white wine?

   (iii) it is either an imported wine *or* a red wine (*or* both)?

   (iv) If only imported wines are selected, what is the probability that it is a red wine?

4.4   Bar charts

A bar chart is a diagram with (vertical) bars that show the number (or percentage) of each outcome of an event. It is a graphic display of the outcomes of a single event.

Note: The height of a bar is proportional to the number of elements associated with each outcome.

Example

There are 50 workers, of whom 16 use only a car, 22 use only a train and 12 use only a bus to travel to work.

(a) Draw a bar chart and give a brief interpretation.

(b) What is the probability that a randomly selected worker:
   • uses a bus to travel to work?
   • travels to work by one mode of public transport (i.e. either a train or a bus)?

Worked solution

The event is: the *mode of transport used*.
The outcomes are: {car, train, bus}
The number (or frequency) for each outcome is:

	Number	**%**
Car	16	32
Train	22	44
Bus	12	24
Total	50	100%

(a) Bar chart (n = 50 workers)

Bar chart (n = 50 workers)

By inspection, more workers use the train (44%) than either a car (32%) or a bus (24%).

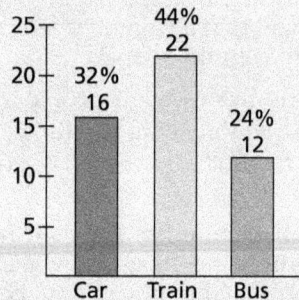

(b) • $P(Bus) = \frac{12}{50} = 0.24$ or 24%
    • $P(Train\ or\ Bus) = \frac{22}{50} + \frac{12}{50} = \frac{34}{50} = 0.68$ or 68% (addition rule)

Note: A worker cannot use multiple modes of transport simultaneously. Therefore these three events {car, train, bus} are mutually exclusive.

Exercise 47

Draw the bar chart for each of the following events and their associated outcomes. Express the numbers (frequencies) for each bar as percentages and interpret the finding.

(a) Event: type of car owned for a sample of 100 car owners
    Outcomes = {Honda = 22; Fiat = 12; Corsa = 38; VW Golf = 28}

(b) Event: city of residence for a sample of 75 holidaymakers to Cape Town
    Outcomes = {Pretoria = 8; Kimberley = 14; Johannesburg = 38; Durban = 15}

(c) Event: preferred grocery store for 150 shoppers
    Outcomes = {Checkers = 57; Spar = 28; Pick 'n Pay = 65}

(d) Event: qualification of a sample of 500 workers
    Outcomes = {Matric = 196; Certificate = 54; Diploma = 94; Degree = 156}

(e) Event: number of children per family in a sample of 250 households
    Outcomes = {0 = 35; 1 = 85; 2 = 64; 3 = 50; 4+ = 16}

4.5 Counting rules: factorials, combinations and permutations

Factorials, combinations and permutations formulae are counting rules used to find the number of outcomes (subsets of elements) out of total number of outcomes (the set) possessing a given characteristic.

Solutions to counting rules rely on an understanding and use of the term *factorial*.

1. Factorials (*n*!)

n! is read as '*n* factorial'.
It is shorthand notation to represent the product of all integers from *n* down to 1.
n! computes the number of distinct orderings (or arrangements) of *n* objects.

Example 1	**Example 2**	**Example 3**
$3! = 3.2.1 = 6$	$5! = 5.4.3.2.1 = 120$	$7! = 7.6.5.4.3.2.1 = 5\,040$

So if three objects can be arranged in any order, then there are a total of $3! = 6$ distinct ways of arranging these three objects.
Similarly, if five objects can be arranged in any order, then there are a total of $5! = 120$ unique orderings of these five objects.
Finally, seven objects can be arranged in 5 040 distinct ways.

Note: $0! = 1$

Note: Factorial answers are always *even numbers* because 2 is always a factor of the answer.

Example 4

In how many different orders can you watch four different DVDs?

Worked solution

$4! = 4.3.2.1 = 24$

2. Multiplication rule of factorials

If there are n_1 objects of type 1, n_2 objects of type 2, n_3 objects of type 3, ... n_k objects of type *k*, then the number of unique orderings (arrangements) of *k* objects is found as follows:

$$(n_1 \times n_2 \times n_3 \times ... \times n_k) \text{ unique orderings of } k \text{ objects}$$

Example

A menu offers 5 starters, 10 main courses and 4 desserts. How many distinct three-course meals consisting of one starter, one main course and one dessert course, can be ordered?

Worked solution

$5 \times 10 \times 4 = 200$ distinct three-course meal orderings

---

Factorials can also be used to compute the number of different orderings for *subsets* of objects drawn from larger sets of objects. These are called combinations and permutations.

3. Combinations ($_nC_r$)

A **combination** is the number of different ways in which a subset of r objects can be selected from a larger group of n objects, where the *ordering* of the objects in each subset *is not important*.

Each possible arrangement is called a combination.

In a combination, the *order* in which the r objects appear in each subset *is not important*.

The formula to determine the number of combinations (order not important) is given by:

$$_nC_r = \frac{n!}{r!(n-r)!}$$

Example

If only three out of five different coloured VW cars can be displayed on a showroom floor, how many different colour combinations can be made?

Worked solution

$$_nC_r = \frac{5!}{3!(5-3)!} = \frac{5!}{3!2!} = \frac{5.4.3.2.1}{(3.2.1)(2.1)} = \frac{120}{12} = 10$$

There are 10 separate colour combinations of three cars from a group of five cars.

4. Permutations ($_nP_r$)

A **permutation** is the number of *distinct ways* of arranging a subset of r objects selected from a larger group of n objects, where the *ordering* of the objects in each subset *is important*.

Each distinct ordering is called a permutation.

In a permutation, the *order* in which the r objects appear in each subset *is important*.

The formula to determine the number of separate permutations is given by:

$$_nP_r = \frac{n!}{(n-r)!}$$

Example

If only three out of five different coloured VW cars can be displayed on a showroom floor at any one time, how many distinct displays (colour ordering is important) can be made up?

Worked solution

$$_nP_r = \frac{5!}{(5-3)!} = \frac{5!}{2!} = \frac{5.4.3.2.1}{2.1} = \frac{120}{2} = 60$$

There are 60 separate ways (permutations) of displaying three out of five cars where colour ordering in which they arranged cars on their showroom floor is important.

Note: For any given n and r values, the number of permutations will *always* be larger than the number of combinations, because each single combination can be re-ordered to give different permutations of the same combination.

Exercise 48

1. Find the value of:

    (a) $_7P_4$ (b) $_8C_0$ (c) $_6P_2$ (d) $_7C_3$

2. (a) If each of six workers can be allocated to separate tasks, how many different ways can the allocations be made?

    (b) There are six levels of shelving in a supermarket. If four brands of corn flakes must each be placed on a separate shelf, in how many different ways can a packer arrange the corn flake brands?

    (c) A student must answer three questions from a choice of eight questions in an examination. How many different groupings of questions can the student answer?

    (d) How many two-element subsets of {1, 2, 3, 4} are there that do not contain the pair of elements 1 and 3?

3. (a) A company employs six engineers and four managers. If a committee is to be created with three engineers and two managers, how many different committees are possible?

    (b) A board of trustees of four members is to be formed from nine candidates. The first person selected will be the convenor, the second person will be the secretary, the third person will be the treasurer and the fourth person will be the ordinary member. How many distinct committees can be formed?

    (c) A building contractor requires a team consisting of one bricklayer, an electrician and a plumber. If there are 12 bricklayers, five electricians and seven plumbers available, how many unique teams can be formed? And what is the probability that a particular combination will be selected?

4.6   Probability trees

A **probability tree** is a diagram that shows the outcomes of *events* that occur in *sequence*.

Example

If a coin is tossed three times, what is the probability of getting only two heads?

Worked solution

For each toss, there are two outcomes (head or tail). The sequence of heads and tails from each toss is shown graphically in the probability tree.
To find the probability of a given *sequence of outcomes* occurring on three tosses, *multiply the individual probabilities* on each path.

So $P(HTT) = \frac{1}{2} \times \frac{1}{2} \times \frac{1}{2} = \frac{1}{8}$. Similarly, for $P(THH) = \frac{1}{2} \times \frac{1}{2} \times \frac{1}{2} = \frac{1}{8}$.

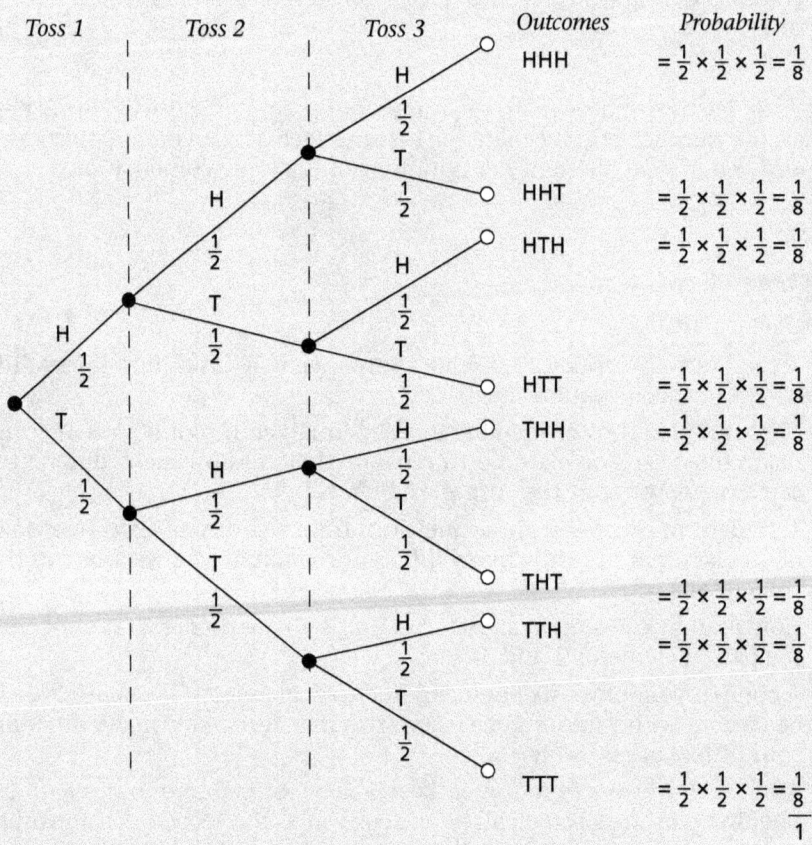

Toss 1	Toss 2	Toss 3	Outcomes	Probability
			HHH	$= \frac{1}{2} \times \frac{1}{2} \times \frac{1}{2} = \frac{1}{8}$
			HHT	$= \frac{1}{2} \times \frac{1}{2} \times \frac{1}{2} = \frac{1}{8}$
			HTH	$= \frac{1}{2} \times \frac{1}{2} \times \frac{1}{2} = \frac{1}{8}$
			HTT	$= \frac{1}{2} \times \frac{1}{2} \times \frac{1}{2} = \frac{1}{8}$
			THH	$= \frac{1}{2} \times \frac{1}{2} \times \frac{1}{2} = \frac{1}{8}$
			THT	$= \frac{1}{2} \times \frac{1}{2} \times \frac{1}{2} = \frac{1}{8}$
			TTH	$= \frac{1}{2} \times \frac{1}{2} \times \frac{1}{2} = \frac{1}{8}$
			TTT	$= \frac{1}{2} \times \frac{1}{2} \times \frac{1}{2} = \frac{1}{8}$
				$\underline{1}$

(branch probabilities on the tree each $\frac{1}{2}$)

Only two heads out of three tosses happens three times.
They are (HHT, HTH, THH).
So P(2 heads out of 3 tosses) = P(HHT) + P(HTH) + P(THH)

$= \frac{1}{8} + \frac{1}{8} + \frac{1}{8} = \frac{3}{8}$ (37.5% chance) (addition rule of probability)

Exercise 49

1. A product consists of two parts, A and B. There is a 1% chance that part A will fail and a $\frac{1}{2}$% chance that part B will fail. What percentage of products is likely to fail if:

   (a) the product fails when both parts are not working

   (b) the product fails when at least one part fails?

2. Refer to the probability tree in the example above.

   (a) What is the probability of getting an identical result in three consecutive tosses of a coin?

   (b) What is the probability that there is at least one tail in each trial of three tosses?

3. Each of 100 balls in a bag is numbered from 1 to 100. If three balls are drawn at random, with replacement after each draw, what is the probability that their sum will be odd?

APPENDIX

A

Pre-revision test

Time: 35 minutes   **No calculators may be used**

This test gives a student the opportunity to benchmark his or her current mathematical knowledge and analytical skills.

All calculations must be done either manually or using mental arithmetic, as this promotes numerical thinking and builds confidence in one's numerical skills.

1. $\frac{(3 \times 0.072)}{0.54} = ?$

    (A) 0.04   (B) 0.3   (C) 0.4

    (D) 0.8   (E) 4.0

2. A car dealer sold x used cars and y new cars during April. If the number of used cars sold was 10 greater than the number of new cars sold, which of the following expresses this relationship?

    (A) $x > 10y$ (B) $x > y + 10$ (C) $x > y - 10$

    (D) $x = y + 10$ (E) $x = y - 10$

3. What is the maximum number of $1\frac{1}{4}$ metre pieces of wire that can be cut from a roll of wire that is 24 metres long?

    (A) 11   (B) 18   (C) 19

    (D) 20   (E) 30

4. What is the value of $\frac{xy}{z}$ when $\frac{x}{z} = \frac{5}{2}$ and $\frac{1}{y} = \frac{1}{10}$?

    (A) 4   (B) 10   (C) 20

    (D) 25   (E) 50

5. $\frac{61.24 \times (0.998)^2}{\sqrt{403}}$

    The expression above is approximately equal to?

    (A) 1   (B) 3   (C) 4

    (D) 5   (E) 6

6. Car x and car y travelled the same 80 kilometre route. If car x took 2 hours and car y traveled at an average speed that was 50% faster than the average speed of car x , how many hours did it take car y to travel the route?

   (A) $\frac{2}{3}$ (B) 1   (C) $1\frac{1}{3}$

   (D) $1\frac{3}{5}$ (E) 3

7. If the numbers $\frac{17}{24}, \frac{1}{2}, \frac{3}{8}, \frac{3}{4}$ and $\frac{9}{16}$ were ordered from greatest to least, the middle number of the resulting sequence would be:

   (A) $\frac{17}{24}$ (B) $\frac{1}{2}$ (C) $\frac{3}{8}$

   (D) $\frac{3}{4}$ (E) $\frac{9}{16}$

8. If a 10% deposit paid toward the purchase of a certain product is R110, how much more remains to be paid?

   (A) R880   (B) R990   (C) R1 000

   (D) R1 100   (E) R1 210

9. Kim purchased n items from a catalogue for R8 each. Postage and handling charges consisted of R3 for the first item and R1 for each additional item.

   Which of the following expressions gives the total amount of Kim's purchase, including postage and handling, in terms of n?

   (A) $8n + 2$ (B) $8n + 4$ (C) $9n + 2$

   (D) $9n + 3$ (E) $9n + 4$

10. $(\sqrt{7} + \sqrt{7})^2 =$

    (A) 98   (B) 49   (C) 28

    (D) 21   (E) 14

11. If the average (arithmetic mean) of the four numbers K, $2K + 3$, $3K - 5$ and $5K + 1$ is 63, what is the value of K?

    (A) 11   (B) $15\frac{3}{4}$ (C) 22

    (D) 23   (E) $25\frac{1}{2}$

12. A rabbit on a controlled diet is fed 300 grams of a mixture of two foods daily, food x and food y. Food x contains 10% protein and food y contains 15% protein. If the rabbit's diet requires exactly 38 grams of protein daily, how many grams of food x are in the mixture?

    (A) 100   (B) 140   (C) 150

    (D) 160   (E) 200

13. The average check-in time for the last five passengers was 4.2 minutes. If the check-in time for four of these passengers was 5.1, 3.8, 4.6 and 3.5 minutes, respectively, what was the check-in time for the 5th passenger?

   (A) 3.7 min   (B) 4 min   (C) 4.2 min
   (D) 4.6 min   (E) 5 min

14. In a survey of 60 sports fans, 45 enjoy soccer, 35 enjoy cricket and 10 do not enjoy either soccer or cricket, but some other sport.

   What percentage of sports fans enjoys both soccer and cricket?

   (A) 20%   (B) 30%   (C) 50%
   (D) 60%   (E) 80%

15. The top of a ladder leans against a vertical wall at 8 m high. If the ladder is 10 m long, how far is the foot of the ladder from the base of the wall?

   (A) 3 m   (B) 4 m   (C) 5 m
   (D) 6 m   (E) 8 m

16. If the roots of a quadratic equation are –2 and $\frac{3}{2}$, what is the quadratic equation?

   (A) $2x^2 + x - 6$ (B) $3x^2 + 2x + 4$ (C) $6x^2 - 4x + 2$
   (D) $x^2 - 3x - 1$ (E) $4x^2 + x - 3$

Post-revision test

Time: 35 minutes   **No calculators may be used**

1. If $\frac{1}{2}$ the air in a tank is removed with each stroke of a vacuum pump, what fraction of the original amount of air has been removed after 4 strokes?

    (A) $\frac{15}{16}$ (B) $\frac{7}{8}$ (C) $\frac{1}{4}$

    (D) $\frac{1}{8}$ (E) $\frac{1}{16}$

2. A restaurant buys fruit in cans containing $3\frac{1}{2}$ cups of fruit each. If the restaurant uses $\frac{1}{2}$ a cup of the fruit in each serving of its fruit dessert, what is the least number of cans needed to prepare 60 servings of the fruit dessert?

    (A) 7   (B) 8   (C) 9

    (D) 10   (E) 12

3. $\left(\frac{1}{5}\right)^2 - \left(\frac{1}{5}\right)\left(\frac{1}{4}\right) = ?$

    (A) $-\frac{1}{20}$ (B) $-\frac{1}{100}$ (C) $\frac{1}{100}$

    (D) $\frac{1}{20}$ (E) $\frac{1}{5}$

4. Mark rented a car for R18 plus R0.10 per km. Colin rented a car for R25 plus R0.05 per km. If each drove d km and each was charged exactly the same amount for the rental, then d equals:

    (A) 100   (B) 120   (C) 135

    (D) 140   (E) 150

5. Machine A produces bolts at a uniform rate of 120 bolts every 40 seconds and machine B produces bolts at a uniform rate of 100 every 20 seconds. If the two machines are run simultaneously, how many seconds will it take for them to produce a total of 200 bolts?

    (A) 22   (B) 25   (C) 28

    (D) 32   (E) 56

6. A fruit salad mixture consists of apples, peaches and grapes in the ratio of 6 : 5 : 2, respectively, by weight. If 78 kg of the mixture is prepared, how many more kg of apples than grapes are in the mixture?

    (A) 30   (B) 24   (C) 18

    (D) 12   (E) 8

7. Which of the following figures has the largest area?
    I. A circle of radius $\sqrt{2}$.
    II. An equilateral triangle whose sides each have length 4.
    III. A triangle whose sides have lengths 3, 4 and 5.

    (A) I   (B) II   (C) III
    (D) I and II   (E) II and III

8. If $\frac{1}{2} + \frac{1}{4} = \frac{x}{15}$, then x is:
    (A) 10   (B) 11.25   (C) 12
    (D) 13.75   (E) 14

9. Three kilograms of grass seed A contains 5% herbicide. A different type of grass seed, grass seed B, which contains 20% herbicide, will be mixed with the three kilograms of grass seed A. How much grass seed B should be added to the three kilograms of grass seed A so that the resulting mixture contains 15% herbicide?
    (A) 3 kg   (B) 3.75kg   (C) 4.5 kg
    (D) 6 kg   (E) 9 kg

10. A car is travelling on a straight highway. At 10 o'clock it passes a truck travelling in the same direction. The truck continues on the highway travelling at 50 km/h, while the car is travelling at a constant speed that is 30% faster than the truck. How far apart are the car and the truck at 2 o'clock?
    (A) 15 km   (B) 30 km   (C) 60 km
    (D) 200 km   (E) 260 km

11. Cereal costs $\frac{1}{3}$ as much as bacon. Bacon costs $\frac{5}{4}$ as much as eggs.
    Eggs cost what percentage of the cost of cereal?
    (A) 41.67%   (B) 80%   (C) 125%
    (D) 167.67%   (E) 240%

12. If $x = 1\,000(0.1)^2$, $y = \frac{1}{(0.1)^2}$, and $z = 100[1 - (1 - 0.1)^2]$, then which one of the following is true?
    (A) $x = y = z$ (B) $x < z < y$ (C) $z < x < y$
    (D) $x < y$ and $x = z$ (E) $x < z$ and $x = y$

13. The median of six consecutive numbers, all of which are multiples of 8, is 52. What is the highest number?

(A) 48   (B) 56   (C) 60

(D) 64   (E) 72

14. In a survey of 60 sports fans, 45 enjoy soccer, 35 enjoy cricket and 10 do not enjoy either soccer or cricket, but some other sport.

What percentage of sports fans enjoys either soccer or cricket or both sports?

(A) 17%   (B) 35%   (C) 56%

(D) 83%   (E) 90%

15. A piece of copper wire 42 cm long is bent into a rectangular shape whose length is twice the width. What is the length of the diagonal of this rectangle?

(A) $7\sqrt{3}$ cm   (B) $5\sqrt{7}$ cm   (C) $10\sqrt{2}$ cm

(D) $7\sqrt{5}$ cm   (E) 21 cm

16. If the roots of a quadratic equation are –2 and k, and the y-intercept of this equation is +4, what is the quadratic equation?

(A) $2x^2 + x - 6$ (B) $x^2 + 4x + 4$ (C) $3x^2 - 4x + 2$

(D) $x^2 - 3$ (E) $x^2 + 2x - 3$

APPENDIX C

Solutions to exercises

CHAPTER 1 Basic arithmetic

Exercise 1

1. (a) $4\sqrt{3}$ (b) $3\sqrt{6}$ (c) $6\sqrt{3}$ (d) $3\sqrt{7}$ (e) $2\sqrt{22}$ (f) $6\sqrt{11}$
    (g) $15\sqrt{3}$ (h) $14\sqrt{3}$ (i) $11\sqrt{6}$ (j) $2\sqrt{15}$ (k) $3\sqrt{5}$

2. (a) $2^4 10^2$ (b) $3^2 10^2$ (c) $2^2 3^1 10^3$ (d) $2^1 3^2 10^4$
    (e) $2^7 10^{-2}$ (f) $3^4 10^{-3}$ (g) $3^2 10^{-4}$ (h) $2^8 10^{-1}$

3. (a) 24 minutes   (b) 2^{10} (c) $2^3\,10^4$

Exercise 2

1. (a) F   (b) T   (c) F   (d) T   (e) F   (f) T
    (g) F   (h) F   (i) F   (j) T   (k) T   (l) F
    (m) T   (n) F   (o) T   (p) F   (q) F   (r) F

2. (a)
(b)
(c)
(d)
(e)
(f)
(g)
(h)
(i)
(j)
(k)
(l)

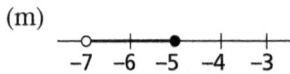

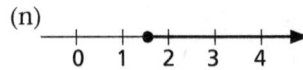

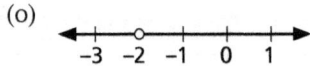

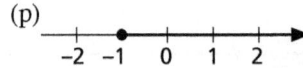

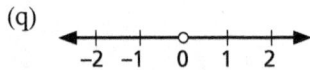

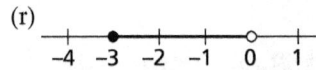

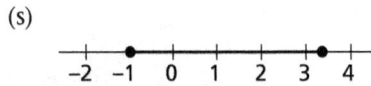

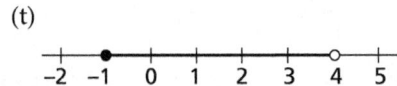

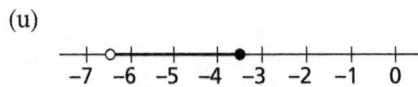

(m)

-7 -6 -5 -4 -3

(n)

0 1 2 3 4

(o)

-3 -2 -1 0 1

(p)

-2 -1 0 1 2

(q)

-2 -1 0 1 2

(r)

-4 -3 -2 -1 0 1

(s)

-2 -1 0 1 2 3 4

(t)

-2 -1 0 1 2 3 4 5

(u)

-7 -6 -5 -4 -3 -2 -1 0

3. (b) and (c) only.
4. (b) only.
5. No (test with negative numbers)
6. (a) F   (b) F   (c) F   (d) F   (e) T
7. 2.5
8. p, t, s, q
9. 14
10. 56
11. (b) only

Exercise 3

1. (a) 5 6 7 8 9 10 11   (b) 28 29 30 31 32 33 34
    (c) 81 82 83 84 85 86 87   (d) 302 303 304 305 306 307 308
2. (a) −12 −11 −10 −9 −8 −7 −6   (b) −33 −32 −31 −30 −29 −28 −27
    (c) −40 −39 −38 −37 −36 −35 −34   (d) −102 −101 −100 −99 −98 −97 −96
3. (a) 17 19 21 23 25 27 29   (b) 3 5 7 9 11 13 15
    (c) −7 −5 −3 −1 +1 +3 +5   (d) −21 −19 −17 −15 −13 −11 −9
4. (a) −2 0 2 4 6 8 10   (b) 18 20 22 24 26 28 30
    (c) −60 −58 −56 −54 −52 −50 −48   (d) −6 −4 −20 24 6
5. (a), (b) and (c)
6. (c) only
7. (b) and (c) only
8. (a) 29   (b) 26   (c) 23
9. (a) 180   (b) 32   (c) 51
10. (a)

Exercise 4

1. (a) 24 (b) 6.25 (c) $\frac{5}{8}$ (d) 5 (e) 13
   (f) 2 (g) 10 (h) 13 (i) 4 (j) –7
   (k) –30 (1) 5
2. A
3. (c)
4. (a) Yes (b) No (c) Yes (d) No

Exercise 5

1. (a) 24 (b) –11 (c) –18 (d) 0 (e) –21
   (f) 1 (g) 19 (h) –15 (i) –11 (j) 14
   (k) –14 (1) –7 (m) 26 (n) –17 (o) –21
   (p) 0 (q) 4 (r) 32
2. $m = 4, n = 1, m + n = 5$
3. (a) 140 (b) –6 (c) –3 (d) 0 (e) 84
   (f) –20 (g) –84 (h) 1 (i) 120 (j) –48
   (k) $\frac{36}{7}$ (1) –20 (m) 96 (n) 12 (o) 0
   (p) 32 (q) 180 (r) 32
4. $m = -10, n = -4, m - n = -6$
5. (a) 26 (b) 9 (c) 39 (d) 11 (e) 22
   (f) 81 (g) 434 (h) 7 (i) 16 (j) 0
   (k) –1 (l) –2 (m) –4 (n) 35 (o) 5
   (p) 20 (q) 3 (r) 0

Exercise 6

1. (a) 84 (b) –54 (c) 135 (d) 72
2. (a) –3 (4) (b) –4 (–12) (c) 9 (7) (d) 9 (3)
3. (a) 1 2 3 4 6 8 12 24 (b) 1 2 5 10
   (c) 1 17 (d) 1 2 4 7 14 28
   (e) 1 2 3 4 5 6 10 12 15 20 30 60 (f) 1 2 3 4 6 8 9 12 18 24 36 72
   (g) 1 2 3 5 6 9 10 15 18 30 45 90 (h) 1 2 5 10 11 22 55 110
   (i) 1 2 5 10 25 50 (j) 1 3 17 51
   (k) 1 37 (l) 1 2 3 4 6 7 12 14 21 28 42 84
4. (a) 5 10 15 20 25 (b) 9 18 27 36 45
   (c) 12 24 36 48 60 (d) 16 32 48 64 80
   (e) 20 40 60 80 100 (f) 50 100 150 200 250
   (g) 250 500 750 1 000 1 250 (h) 600 1 200 1 800 2 400 3 000

5. (a) 1 3 11 33   (b) 1 3 13 39   (c) 1 2 3 6 9 18
    (d) 1 3 9 27   (e) 1 31 prime   (f) 1 3 17 51
    (g) 1 53 prime   (h) 1 43 prime   (i) 1 3 41 123
    (j) 1 3 7 9 21 63   (k) 1 3 9 13 39 117   (l) 1 2 11 22

6. (a) $2 \times 2 \times 2 \times 2$ (b) $2 \times 3 \times 5$ (c) 3×11
    (d) $2 \times 2 \times 2 \times 7$ (e) 19   (f) $3 \times 3 \times 5$
    (g) $2 \times 3 \times 13$ (h) $2 \times 2 \times 3 \times 11$ (i) $2 \times 2 \times 7$
    (j) 3×13 (k) $2 \times 3 \times 3 \times 3$ (l) $3 \times 3 \times 7$

7. (a) F   (b) F   (c) T
8. (e) 23
9. 71
10. $x + y = 12$
11. Only (d)
12. Only (b)
13. (a), (b) and (d) only
14. 7
15. Only (e)
16. $3 \times 17 = 51$
17. Only (d)
18. 28

Exercise 7

1. (a) 6   (b) 14   (c) 21   (d) 15   (e) 4
    (f) 14   (g) 33   (h) 18   (i) 25   (j) 16   (k) 18
2. (a) 40   (b) 48   (c) 36   (d) 90   (e) 420
    (f) 30   (g) 105   (h) 80   (i) 30   (j) 210
3. (a) T   (b) F [72]   (c) F [26]   (d) F [16]
4. (b) and (c)

Exercise 8

1. (a) $\frac{5}{24}$ (b) $\frac{29}{36}$ (c) $1\frac{11}{21}$ (d) $4\frac{9}{20}$ (e) $-1\frac{11}{24}$
2. (a) $\frac{7}{12}$ (b) 1   (c) $\frac{4}{7}$ (d) $3\frac{9}{10}$ (e) $1\frac{5}{12}$
3. (a) $1\frac{5}{16}$ (b) $\frac{25}{81}$ (c) $1\frac{2}{7}$ (d) $2\frac{17}{24}$ (e) $\frac{16}{51}$
4. (a) $\frac{1}{17}$ (b) $-2\frac{4}{13}$ (c) $1\frac{7}{33}$ (d) 7   (e) $8\frac{1}{6}$ (f) $1\frac{4}{45}$
5. 17

Exercise 9

1. (a) 0.04 0.38 0.4   (b) 0.8 0.809 0.81
   (c) 0.0999 0.99 0.998 1   (d) 3.99 4.0 4.024 4.03
2. (a) T   (b) F   (c) T   (d) T   (e) F (they are equal)
3. (d) $\frac{8}{11}$
4. −0.01

Exercise 10

1. (a) 6.4   (b) 72 750   (c) 812.5   (d) 0.045   (e) 787.7
   (f) 0.036   (g) 1.9784   (h) 0.000082   (i) 12.72394   (j) 0.080003
2. (a) 0.375   (b) 0.5555   (c) 0.30   (d) 0.41667   (e) 0.384615
   (f) 0.8333   (g) 0.06   (h) 0.4375
3. (a) 27.714   (b) 21.315   (c) 20.125   (d) 0.0625   (e) 85.47
   (f) 1.36   (g) 18.904   (h) 6.965   (i) 15.936   (j) 69.993
4. 0.07309

Exercise 11

1. (a) 5.2   (b) 0.0312   (c) 0.0532   (d) 0.000816   (e) 0.03003
   (f) 1.245   (g) 20.53   (h) 0.3168   (i) 0.00832   (j) 0.2331
2. (a) 78   (b) 8.7   (c) 41.1   (d) 5 100   (e) 300.3
   (f) 0.055   (g) 82.1   (h) 177   (i) 22.07   (j) 25 900
3. (a) 8.38   (b) 7.4   (c) 61.05   (d) 0.45   (e) 0.0236
   (f) 0.072   (g) 0.02111   (h) 0.00085   (i) 0.235864
4. (a) 4.35   (b) 9.16   (c) 1.008   (d) 20.65   (e) 10.2
   (f) 2.95   (g) 0.62   (h) 2.8   (i) 2.658   (j) 34 640
   (k) 66.44   (l) 0.0048   (m) 6.65   (n) 0.00584   (o) 0.009
   (p) 3.45   (q) 0.01998   (r) 0.2331   (s) 9.8   (t) 65.14

Exercise 12

1. (a) 27   (b) 1   (c) 216   (d) 10 000   (e) 64
   (f) −64   (g) −1   (h) 625   (i) 10 000   (j) 1
2. (a) $\frac{1}{27}$ (b) 1   (c) $\frac{1}{256}$ (d) $\frac{1}{10\,000}$ (e) $\frac{1}{1\,024}$
   (f) −1   (g) $-\frac{1}{1\,000}$ (h) $\frac{1}{25}$ (i) $\frac{1}{1\,000\,000}$
3. (a) 6^9 (b) 8^4 (c) $\frac{1}{3}$ (d) 1   (e) 10^9
   (f) 1   (g) 4^3 (h) 7^4 (i) $\frac{1}{9^2}$ (j) $\frac{1}{2^5}$
   (k) 3^2 (l) 6   (m) $\frac{1}{5}$

4. (a) $\frac{1}{6}$ (b) $\frac{1}{8^2}$ (c) $\frac{1}{3^7}$ (d) $\frac{1}{5}$ (e) 10^5

(f) 1 (g) 1 (h) $\frac{1}{8^5}$ (i) 5^4 (j) 10^4

(k) 2 (l) 1 (m) $\frac{1}{6}$

5. (a) 2^{20} (b) $\frac{1}{8^6}$ (c) $\frac{1}{3^4}$ (d) 5^6

(e) 10^{10} (f) 1 (g) 1

6. (a) T (b) F (c) T (d) F (e) F

(f) F (g) F (h) T (i) T (j) T

(k) F (l) F (m) F (n) F (o) F

(p) F (q) F (r) T (s) T (t) T

7. (a) $3^3 5^3$ (b) $2^2 11^2$ (c) $(-2)^3 3^3$ (d) $3^2 13^2$

8. (a) 20^3 (b) 21^2 (c) 3^2 (d) 3^3

(e) 20^2 (f) 30^4 (g) 14^3

9. 9

10. 2^5

11. 6^4

Exercise 13

1. (a) 87.5% (b) 55.56% (c) 41.667% (d) 43.75%
    (e) 26.67% (f) 8.33% (g) 35%
2. (a) 56% (b) 70% (c) 7% (d) 42.5% (e) 22%
3. (a) 3 (b) R36 (c) 96 kg (d) 105 g (e) R37.50
4. (a) 25% (b) 30% ; 45% ; 25% (c) R450 (d) R2 000 (e) R10 600
5. (a) 12.5% (b) 56% (c) 25% (d) 10%
6. (a) 33.33% (b) 37.5% (c) 25%
7. (a) 20% (b) 12.5% (c) 5% (d) 33.33%
8. (a) 25% (b) 30% (c) 20% (d) 25%
9. (a) R36 (b) R138 (c) 105 kg (d) 750 cm (e) R73.50
10. (a) 32 kg (b) R180 (c) 45 cm (d) 10.5 cm (e) R100
11. (a) 56 kg (b) R200 (c) 600 km (d) 300 g (e) R51
    (f) 72 kg (g) R128 (h) 80 km
12. (a) R70 (b) 120 kg (c) 50 km
    (d) 400 g (e) 200 km (f) R160
13. (a) R90 (b) 160 kg (c) 50 km
    (d) 600 g (e) R400 (f) 96 m
14. 64%
15. 96 learners

Exercise 14

1. (a) R136   (b) R736   (c) R68   (d) R440
   (e) R253   (f) 199.8 kPa   (g) R504   (h) R17 500
2. (a) 19%   (b) 40%   (c) 32.25%   (d) 20%
3. (a) R400   (b) R400 000
   (c) 12 litres per 100 km   (d) R500

Exercise 15

1. (a) R630; R2 730   (b) R3 840; R10 240   (c) Rl 440; R10 440
2. (a) R360; Rl 560   (b) R6 000; R14 000   (c) R720; R4 720

Exercise 16

1. (a) R420 ; R2 420   (b) R364 ; R864   (c) R9.70 ; R59.70
   (d) R485.95 ; R1 485.95   (e) R105.83 ; R355.83   (f) R215.79 ; R1 015.79
2. (a) R108; R23.10; R385.20 (b) 120   (c) 1 776
   (d) 40%; 25%; 30%; 5%   (e) 16.67%   (f) 12.5%
   (g) 11.11%   (h) 16.67%   (i) R225; R202.50; 12.5%
   (j) R468   (k) Rl 386   (l) 33.33%; 60%
   (m) R200; 50%   (n) 42 000 kg   (o) 20.11%

Exercise 17

1. (a) R45; R75   (b) Rl 400; R4 900   (c) 104; 156; 260
   (d) R132; R88; R220; R220   (e) 80 g   (f) R450
   (g) 56 hours   (h) 16 kg
2. (a) R32   (b) R32; R64   (c) R96; R72
   (d) R100; R160; R200   (e) R18; R27; R9
3. (a) 8   (b) 9   (c) R200   (d) 30 g; 40 g
   (e) 80   (f) 70   (g) R6 000   (h) 44 men
   (i) 22 men   (j) $\frac{5}{8}$ (k) 2 400 votes

Exercise 18

1. (a) R132   (b) 100   (c) 140 min (2 h 20 min)
   (d) R378   (e) 1 400   (f) 105 min (1 h 45 min)
   (g) R7.20   (h) 200 ℓ (i) R2 670; 11
   (j) R4 200   (k) 1 800 bottles
2. (a) 8 hours   (b) 14 hours   (c) 15 days

Exercise 19

1. (a) 82 km/h (b) 2 800 km (c) 0.142857 cm/day (d) 500 m
   (e) 0.6 h (36 min) (f) 0.4 h (24 min) (g) 68.6 km/h (h) $1\frac{2}{3}$ km/h

Exercise 20

1. (a) 24 hours (b) 25 seconds
   (c) 90 km (d) $\frac{47}{180}$; 52.2% ; 3.83 h (3 h 50 min)
   (e) 1 440 ; 43 200 (f) $\frac{9}{20}$
   (g) $5\frac{1}{3}$ hours (h) $1\frac{1}{4}$ hours
   (i) 10 hours (j) $6\frac{2}{3}$ hours

Exercise 21

1. (a) 50.6 (b) 2.34 (c) 31.5 (d) 210 (e) 150
   (f) 396.9
2. (a) no (33) (b) yes (c) yes (d) no (69) (e) no (123)
   (f) no (339) (g) yes (h) no (1 233)
3. (a) yes (b) no (78) (c) no (270) (d) no (672) (e) yes
   (f) no (306 or 312) (g) yes (h) yes
4. (a) yes (b) yes (c) no (234) (d) no (666) (e) yes
   (f) yes (g) yes (h) no (1 935)
5. (a) yes (b) no (88) (c) yes (d) no (352) (e) yes
   (f) no (924) (g) yes (h) yes
6. (a) $\frac{27}{5}$ (b) $\frac{60}{10}$ (c) $\frac{13}{2}$ (d) $\frac{18}{1}$ (e) $\frac{707}{7}$
   (f) $\frac{1\,078}{100}$ (g) $\frac{21}{3}$ (h) $\frac{2\,364}{100}$ (i) 30×100 (j) 2.1×5
   (k) 25×20 (l) 974×0.1 (m) $19 - 14$ (n) $207 + 19$ (o) $664 \times \frac{1}{2}$
7. (a) 324 (b) 484 (c) 1 225 (d) 3 969 (e) 3 025
   (f) 9 409

CHAPTER 2 Fundamental algebra

Exercise 22

(a) $24 + x + z$ (b) $\frac{(w + 3)}{(w + m + 5)}$ (c) R$\frac{3s}{200}$ (d) $\frac{y}{x}$ minutes
(e) R$\frac{(y - x)}{2}$ (f) $\frac{11x}{y}$ seconds (g) $\frac{3y}{2} - 6$ years old (h) R$\frac{3x}{80}$
(i) $\frac{(y - x)}{2}$ (j) $\frac{3q}{200}$

Exercise 23

1. (a) $8x - 8y$ (b) $10q - 3r$ (c) $7a - 11b + 2$
   (d) $5d$ (e) $2a^2 - 4ab + 7b^2$ (f) $9pq - 2q^2 - 6p^2$
   (g) $x^2y + 2xy^2 + 6x^2$ (h) $-13abc + 6ab$

2. (a) $7m + 12$ (b) $22t - 25$ (c) $-3r + 15$
   (d) $-14v + 2$ (e) $-24x + 17$ (f) $14 - 19m$
   (g) $1 - 9y$ (h) $2x + 2y + 4$ (i) $2a + 3a^2$
   (j) $7y^2 - 8y - 6$ (k) $3x - 8y + 4$ (l) $8a - 12$
   (m) $-8x + 7y$

3. (a) $12m^2 + 21m + 9$ (b) $8t^2 + 10t - 25$ (c) $12r^2 - 29r + 15$
   (d) $25c^2 - d^2$ (diff in sq)   (e) $-6y^2 + 5y + 4$ (f) $4x^2 - 5xy - 6y^2$
   (g) $72p^2 - 25pq + 2q^2$ (h) $2a^2 - 6ab + 4b^2$ (i) $b^2 - 4$ (diff in sq)
   (j) $49m^2 - 16$ (diff in sq)   (k) $5v^2 + 19v + 12$ (l) $4p^2 - 13pq + 3q^2$
   (m) $y^2 + 6y + 9$ (n) $4p^2 - 13pq + 3q^2$ (o) $x^2 + x - 6$
   (p) $b^2 - 2b - 8$ (q) $4k^2 - 9$ (diff in sq)   (r) $6d^2 + d - 2$
   (s) $x^2 + 4x + 4$ (t) $16a^2 - 40ab + 25b^2$

4. (a) 22   (b) 12   (c) 4   (d) 56
   (e) 18   (f) 2   (g) 9   (h) 0
   (i) 225   (j) 225   (k) 232   (l) -69
   (m) $4\frac{3}{16}$ (n) $-5\frac{1}{3}$ (o) 16   (p) $-131\frac{1}{5}$
   (q) -78 (r) 55   (s) $\frac{63}{64}$ (t) -5 ; 64 ; -4
   (u) 34 ; 69 ; 12

Exercise 24

1. (a) $\dfrac{(2x^2 + 5y)}{8x}$ (b) $\dfrac{(15z - 16y^2)}{10yz}$ (c) $\dfrac{(14ac - 27b^2)}{12bc}$ (d) $\dfrac{(50v + 3tu^2)}{tv}$
   (e) $\dfrac{17a}{15}$ (f) $\dfrac{5x}{12}$ (g) $\dfrac{(28b + 27a^2)}{24ab^2}$ (h) $\dfrac{(10 + 5m - 14m^3)}{10m^2}$

2. (a) $\dfrac{3n}{m}$ (b) $3uv$ (c) $\dfrac{4}{3xy}$ (d) $\dfrac{10p^2}{q^4}$
   (e) $\dfrac{xy^3}{6}$ (f) $\dfrac{1}{6a^2b}$ (g) $\dfrac{6}{5x}$ (h) $\dfrac{225}{8b^4}$
   (i) $6c$ (j) $\dfrac{3}{2}$ (k) $\dfrac{4}{5n}$ (l) $\dfrac{3r(p - q)}{8}$
   (m) $\dfrac{5x^2}{[3y(2x + y)]}$ (n) $\dfrac{x}{4}$

Exercise 25

1. (a) q^9 (b) 3^5 (c) $\dfrac{1}{y^8}$ (d) $12a^2$
   (e) 56   (f) $\dfrac{m^2}{p}$ (g) z^4 (h) m^3

(i) 5^2 (j) $5p^3$ (k) $9r^4$ (l) $\frac{a^5}{5}$

(m) $\frac{1}{3}$ (n) $\frac{y^5}{x^5}$ (o) $\frac{a^5}{b^8}$ (p) a^{18}

(q) $\frac{1}{x^6}$ (r) $27y^9$ (s) $1^{32} = 1$ (t) 10^6

(u) a^9b^{15} (v) $32n^{10}$ (w) $\frac{1}{16}y^8$ (x) $\frac{16}{81}q^{20}$

Exercise 26

1. (a) $a = \frac{(-2b + 5c)}{2}$ (b) $y = -2z$ (c) $y = \frac{1}{(m-1)}$ (d) $x = \frac{5}{(y+5)}$

2. (a) $c = 11$ (b) $p = -6$ (c) $y = 5$ (d) $a = -10$

    (e) $c = 10$ (f) $x = 12$ (g) $x = \frac{5}{4}$ (h) $a = 5$

    (i) $x = \frac{3}{2}$ (j) $x = \frac{8}{3}$ (k) $x = 3$ (l) $x = 3$

    (m) $x = 2$ (n) $x = -4$ (o) $x = 10$ (p) $x = 3$

    (q) $x = 1$ (r) $x = 4$ (s) $x = 4$ (t) $x = -3$

    (u) $p = -3$ (v) $x = 2$ (w) $x = -2$ (x) $x = 12$

    (y) $c = 2$ (z) $a = \frac{8}{17}$

3. (a) $x = 3$ (b) $x = 5$ (c) $x = 2$ (d) $x = 6$

4. 21   5. -35   6. 16   7. 5

8. 15   9. $k = 14$ 10. $f(x) = \frac{x}{2}$ and $f(x) = |x|$

11. $f(x) = x^2(1 - x)^2$

Exercise 27

1. (a) $2m(3 + n - 9m^2)$ (b) $2pqr(3pr^2 - 1)$ (c) $-4x^2(2xy^2 + 3 + 4x^2y^3)$

    (d) $6ab(a^2bc - 2b + ac)$

2. (a) $(2x + 5)(2x - 5)$ (b) $\left(x + \frac{1}{2}\right)\left(x - \frac{1}{2}\right)$ (c) $\left(b + \frac{5}{6}\right)\left(b - \frac{5}{6}\right)$

    (d) $(5m - 4)(5m + 4)$ (e) $\left(3a - \frac{1}{4}\right)\left(3a + \frac{1}{4}\right)$ (f) $(7x - 3y)(7x + 3y)$

3. (a) 400   (b) 1 440   (c) 720

    (d) 0.2   (e) 0.8   (f) 30

    (g) 39   (h) 17.5   (i) 0.64

4. (a) $(y + 2)(y + 8)$ (b) $(a + 3)(a - 8)$ (c) $(2x + 5)(3x - 1)$ (d) $(r - 6)(2r - 7)$

    (e) $(t - 9)(t - 2)$ (f) $(2a + 3)(5a - 4)$ (g) $(3x - 4)(5x + 1)$ (h) $(y + 3)(2y - 7)$

    (i) $(3c + 1)(2c - 5)$ (j) $(2x - 3)(3x + 5)$ (k) $(2x + 1)(x - 1)$ (l) $(4x + 3)(3x - 4)$

    (m) $(x - 7)(x + 2)$ (n) $(4x - 2)(2x - 4)$

5. (a) $y = -2$ or $y = -8$ (b) $a = -3$ or $a = 8$ (c) $x = -\frac{5}{2}$ or $x = \frac{1}{3}$

    (d) $r = 6$ or $r = \frac{7}{2}$ (e) $t = 9$ or $t = 2$ (f) $a = -\frac{3}{2}$ or $a = \frac{4}{5}$

    (g) $x = \frac{4}{3}$ or $x = -\frac{1}{5}$ (h) $y = -3$ or $y = \frac{7}{2}$ (i) $c = -\frac{1}{3}$ or $c = \frac{5}{2}$

    (j) $x = \frac{3}{2}$ or $x = -\frac{5}{3}$ (k) $x = -\frac{1}{2}$ or $x = 1$ (l) $x = -\frac{3}{4}$ or $x = \frac{4}{3}$

    (m) $x = 7$ or $x = -2$ (n) $x = \frac{1}{2}$ or $x = 2$

6. Only (d)

7. (a) $y = x^2 + x - 6$　　(b) $y = 8x^2 - 6x + 1$　　(c) $y = 2x^2 + 7x - 4$

　　(d) $y = 5x^2 - 7x - 6$　　(e) $y = x^2 - 7x + 10$　　(f) $y = 15x^2 + 19x + 6$

　　(g) $y = 10x^2 + 7x - 12$

8. $x = -7$

9. $k = -2$

Exercise 28

1. (a) IV　　　　　　　　　　(b) II　　　　　　　　　　(c) I

　　(d) between III and IV　　(e) IV　　　　　　　　　(f) between I and IV

　　(g) at the origin　　　　　(h) between I and II　　　(i) III

2. (a)　　　　　　　　　　　　　　　　　　(b)

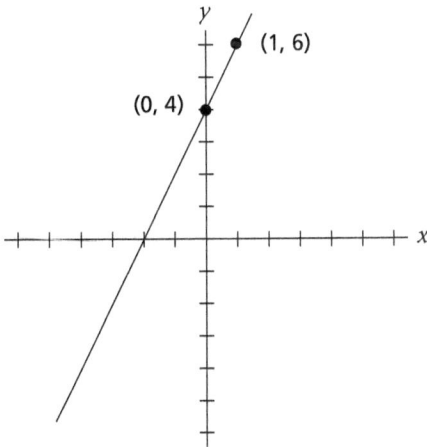

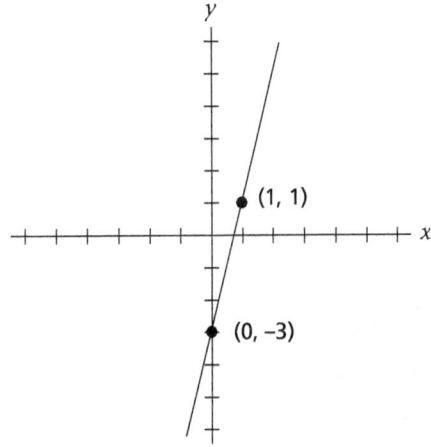

(c)　　　　　　　　　　　　　　　　　　(d)

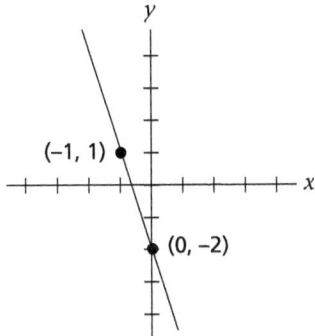

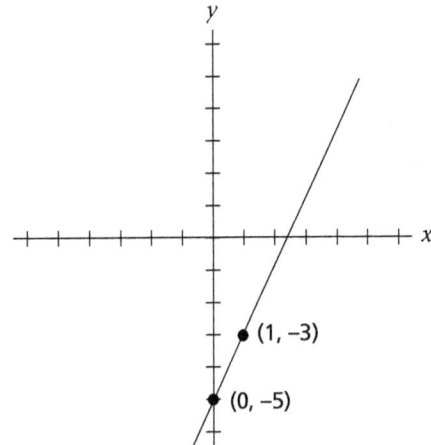

(e)

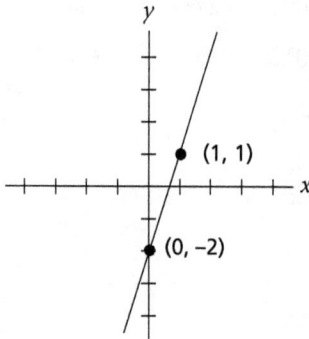

(f)

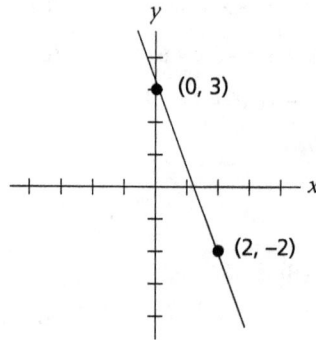

(g)

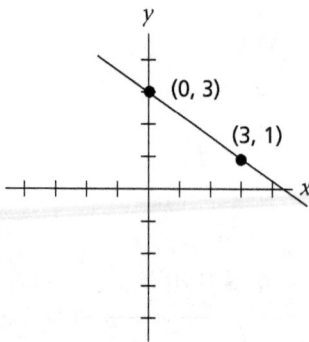

(h)

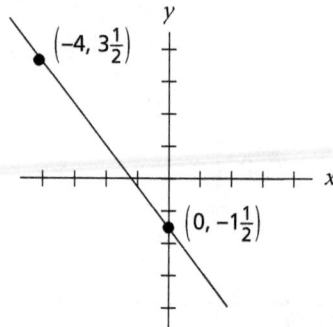

Exercise 29

1. (a) $y = 6x - 9$ (b) $y = -x + 9$ (c) $y = 7$

    (d) $y = \frac{1}{4}x + \frac{13}{24}$ (e) $y = x + 5$ (f) $y = \frac{1}{2}x - 1\frac{1}{2}$

    (g) $y = -\frac{18}{25}x + \frac{11}{50}$

2. (a) 10 (b) 5 (c) 13

    (d) 10 (e) $\sqrt{20}$ (f) 5

3. (a) perpendicular (b) parallel (c) perpendicular

    (d) parallel (e) neither (f) perpendicular

    (g) perpendicular

4. $y = \frac{1}{2}x - 2$

Exercise 30

1. (a) y-intercept = 16 (b) y-intercept = -24 (c) y-intercept = -5

    roots = -2 and -8 roots = -3 and 8 roots = $-2\frac{1}{2}$ and $\frac{1}{3}$

    AS = -5 AS = $2\frac{1}{2}$ AS = $-1\frac{1}{12}$

    TP = (-5, -9) TP = $\left(2\frac{1}{2}, -30\frac{1}{4}\right)$ TP = $\left(-1\frac{1}{12}, -12\frac{1}{24}\right)$

(d) y-intercept = 42
   roots = $3\frac{1}{2}$ and 6
   AS = $4\frac{3}{4}$
   TP = $\left(4\frac{3}{4}, -3\frac{1}{8}\right)$

(e) y-intercept = 18
   roots = 2 and 9
   AS = $5\frac{1}{2}$
   TP = $\left(5\frac{1}{2}, -12\frac{1}{4}\right)$

(f) y-intercept = –12
   roots = $-1\frac{1}{2}$ and $\frac{4}{5}$
   AS = $-\frac{7}{20}$
   TP = $\left(-\frac{7}{20}, -13\frac{9}{40}\right)$

(g) y-intercept = – 4
   roots = $-\frac{1}{5}$ and $1\frac{1}{3}$
   AS = $\frac{17}{30}$
   TP = $\left(\frac{17}{30}, -8\frac{49}{60}\right)$

(h) y-intercept = –21
   roots = 3 and $3\frac{1}{2}$
   AS = $\frac{1}{4}$
   TP = $\left(\frac{1}{4}, -21\frac{1}{8}\right)$

(i) y-intercept = –5
   roots = $-\frac{1}{3}$ and $2\frac{1}{2}$
   AS = $1\frac{1}{12}$
   TP = $\left(1\frac{1}{12}, -12\frac{1}{24}\right)$

(j) y-intercept = –4
   roots = $-\frac{1}{2}$ and 2
   AS = $\frac{3}{4}$
   TP = $\left(\frac{3}{4}, -6\frac{1}{4}\right)$

(k) y-intercept = –1
   roots = $-\frac{1}{2}$ and 1
   AS = $\frac{1}{4}$
   TP = $\left(\frac{1}{4}, -1\frac{1}{8}\right)$

(l) y-intercept = –25
   roots = 5 and $-2\frac{1}{2}$
   AS = $1\frac{1}{4}$
   TP = $\left(1\frac{1}{4}, -28\frac{1}{8}\right)$

(m) y-intercept = –14
   roots = –2 and 7
   AS = $2\frac{1}{2}$
   TP = $\left(2\frac{1}{2}, -20\frac{1}{4}\right)$

Exercise 31

1. (a) $x = 5$ $y = 3$
   (b) $a = 1$ $b = 4$
   (c) $m = 2$ $n = 3$
   (d) $x = 3$ $y = 4$
   (e) $m = 2$ $n = 4$
   (f) $p = 3$ $q = 4$
   (g) $b = 3$ $c = 2$
   (h) $x = 3$ $y = 2$
   (i) $x = 1$ $y = 2$
   (j) $t = 2$ $u = 2$
   (k) $a = 2$ $b = 5$
   (l) $p = 3$ $q = 2$
   (m) $x = 4$ $y = 0$

Exercise 32

1. (a) 8   (b) 5   (c) 48   (d) $4\frac{1}{2}$ (e) 12
   (f) 4   (g) 57   (h) 17   (i) 26 kg   (j) 40 45 50 55
2. (a) 48   (b) $x + y = 375$; $2x - y = 150$ $x = 175$ kg; $y = 200$ kg
   (c) $x + y = 24$; $4x + 2.8y = 84$ x (pens) = 14; y (pencils) = 10
   (d) 140 km each
   (e) R6
   (f) $t + 14 = m$; $m + 10 = 2(t + 10)$ m (currently) = 18 years; m (in 5 years) = 23 years
   (g) $\frac{1}{8}x = \frac{1}{4}\left(\frac{1}{3}x + 40\right)$; $x = 240$ students
   (h) 64   (i) 6   (j) 16   (k) 70   (l) R85
   (m) 20 managers   (n) 240 km   (o) 3 : 2
3. (a) 20   (b) 15th floor   (c) 24 km   (d) 37.5 km   (e) 3 km/h

Exercise 33

(a) >   (b) <   (c) <   (d) <   (e) <   (f) >

(g) <   (h) >   (i) >   (j) >   (k) =   (l) <

(m) <   (n) >   (o) =   (p) =   (q) >   (r) =

(s) =   (t) >   (u) >

Exercise 34

1. (a) $x > 3$ (b) $x \geq 5$ (c) $x < 4$ (d) $x \leq 6$

    (e) $p > 8$ (f) $p \leq 9$ (g) $x > 4$ (h) $x \geq -5$

    (i) $x > -12$ (j) $y \geq 5$ (k) $y \leq 6$ (l) $q < -8$

    (m) $y < 8$ (n) $x > 5$ (o) $x \leq 4$ (p) $p > -2$

    (q) $x \leq 4$ (r) $y > 3$

2. (a) $x = \{1, 4, 9, 16, 25, 36\}$ (b) $x = \{1, 3, 5\}$

    (c) $p = \{6, 12, 18, 24, 30\}$ (d) $x = \{13, 14, 15\}$

    (e) $m = \{1, 9, 25, 49\}$ (f) $y = \{5, 6, 7, 8\}$

    (g) $x = \{-4, -3, -2, -1, 0\}$ (h) $y = \{-6, -5, -4\}$

    (i) $p = \{-16, -14, -12, -10\}$ (j) $y = \{-6, -4, -2\}$

Exercise 35

1. (a)

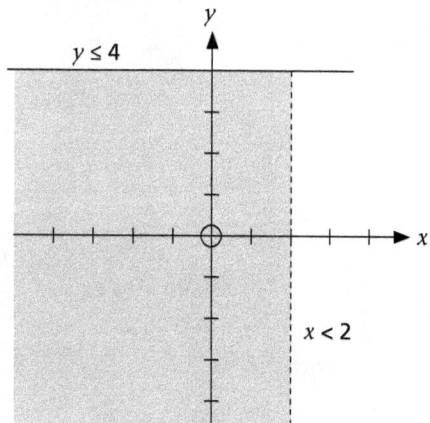

(b)

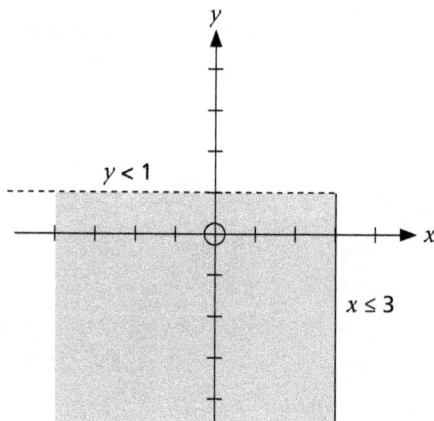

(c)

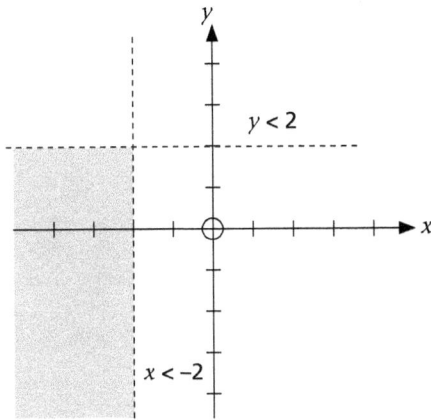

$y < 2$

$x < -2$

(d)

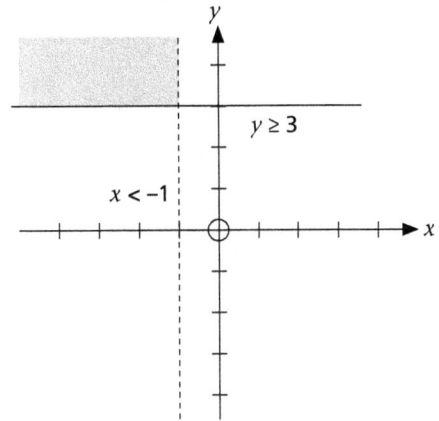

$y \geq 3$

$x < -1$

(e)

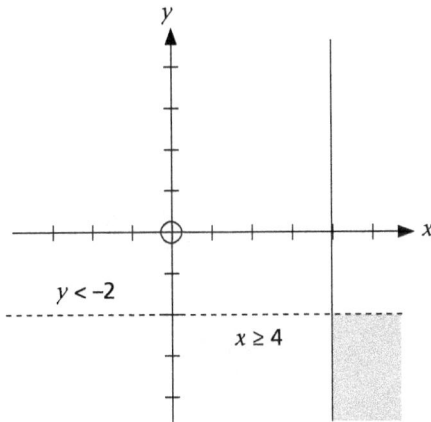

$y < -2$

$x \geq 4$

(f)

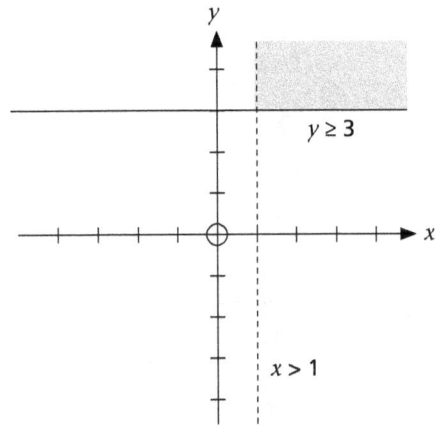

$y \geq 3$

$x > 1$

(g)

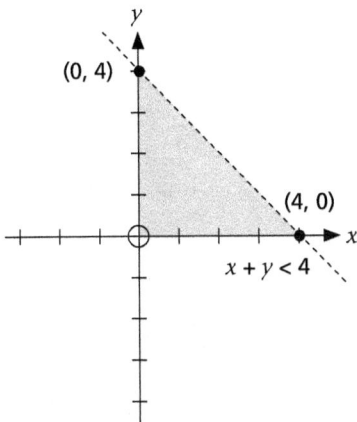

(0, 4)

(4, 0)

$x + y < 4$

(h)

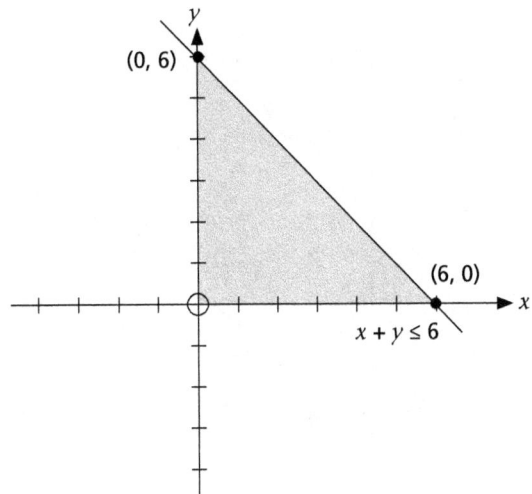

(0, 6)

(6, 0)

$x + y \leq 6$

(i)

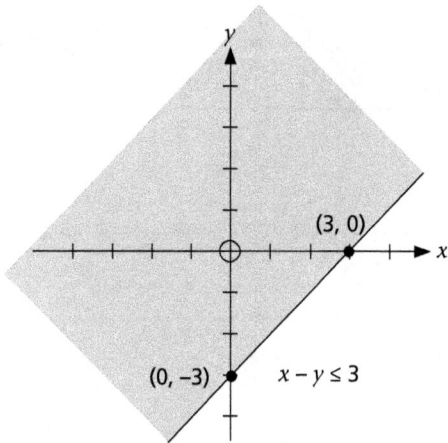

(3, 0)

(0, −3) $x - y \leq 3$

(j)

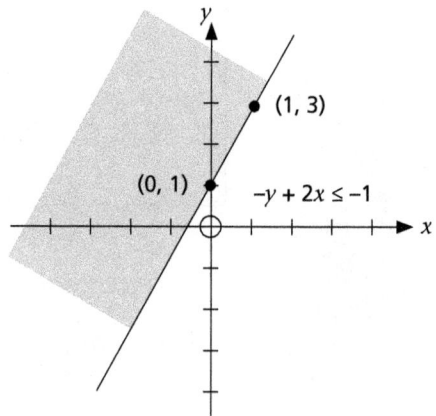

(1, 3)

(0, 1) $-y + 2x \leq -1$

(k)

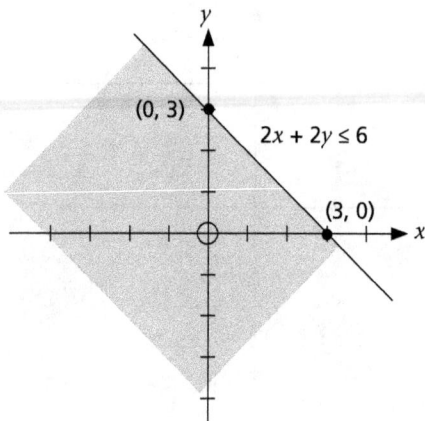

(0, 3) $2x + 2y \leq 6$

(3, 0)

(l)

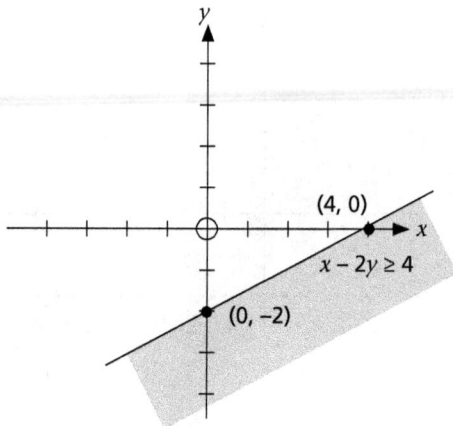

(4, 0)

$x - 2y \geq 4$

(0, −2)

(m)

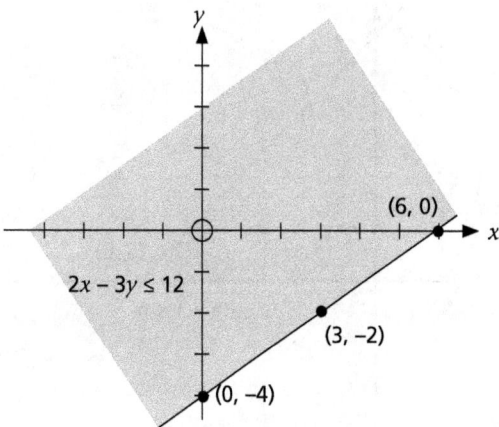

(6, 0)

$2x - 3y \leq 12$

(3, −2)

(0, −4)

(n)

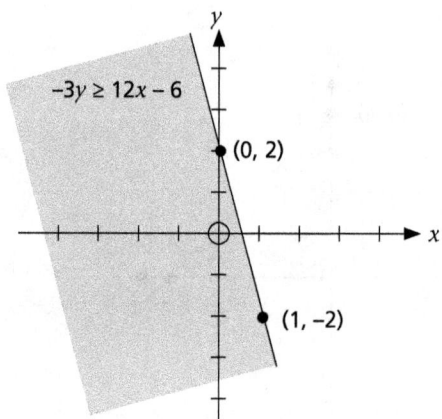

$-3y \geq 12x - 6$

(0, 2)

(1, −2)

2. (a) No

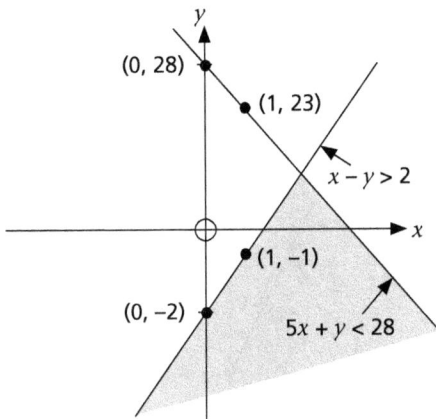

(b) No

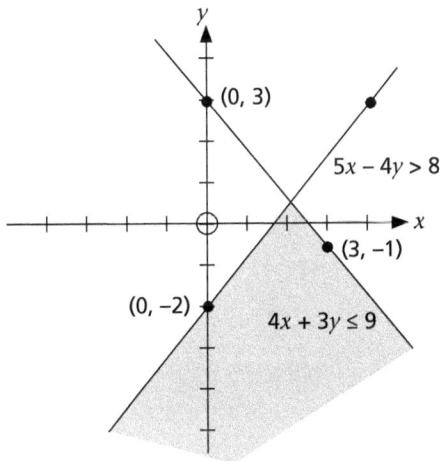

(c) No

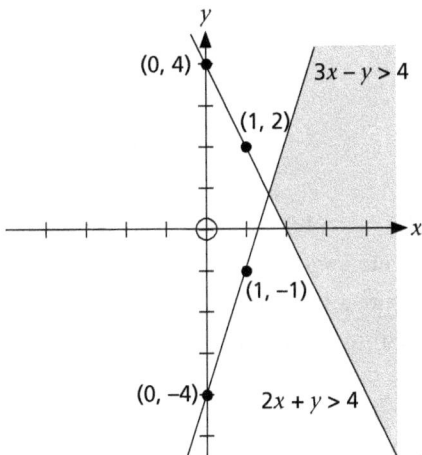

(d) No

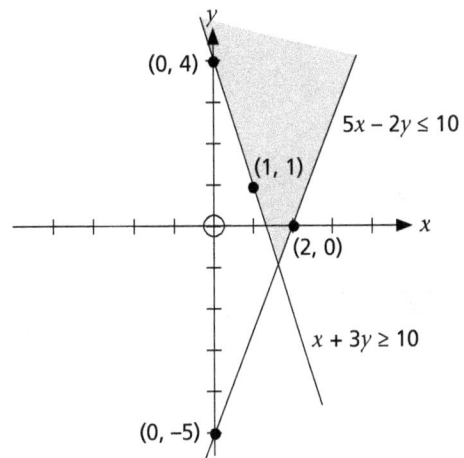

(e) No

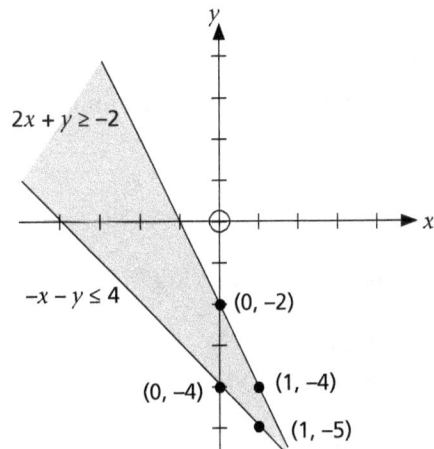

(f) No

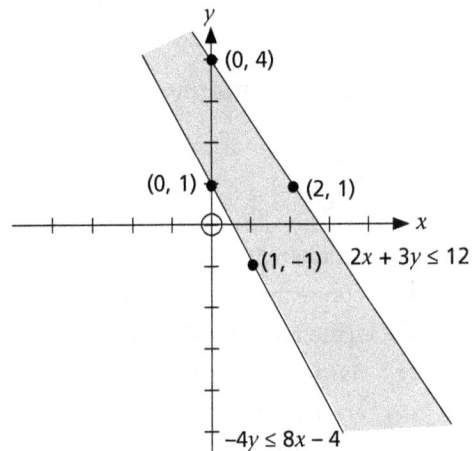

(g) No   (h) No

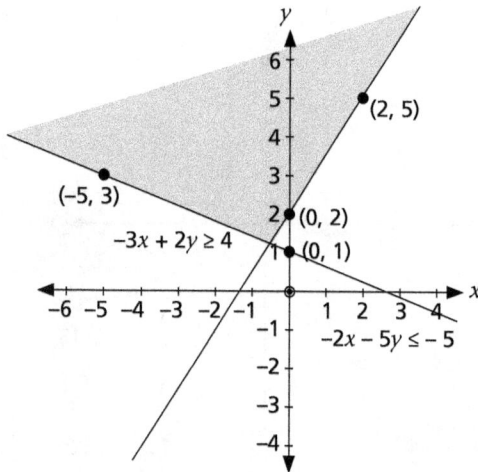

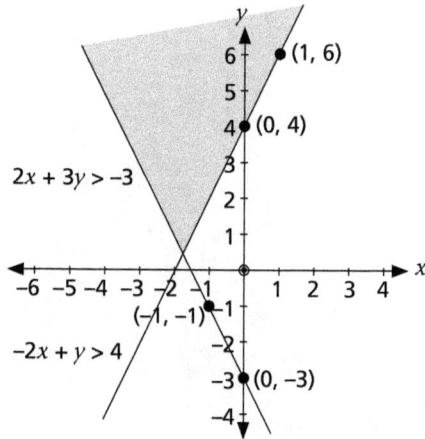

3. IV   4. I and IV

Exercise 36

1. 1 000 units   2. 640 units   3. 700 units

4.

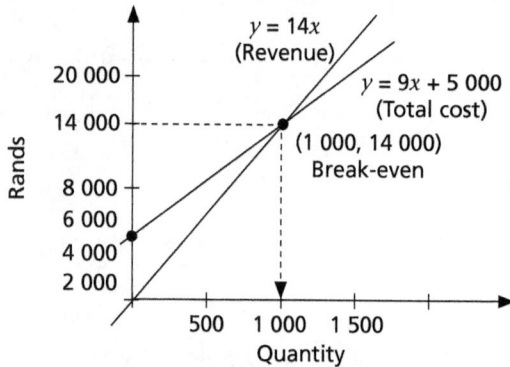

CHAPTER 3 Geometry

Exercise 37

1. (a) C   (b) F   (c) A   (d) D   (e) C   (f) B

2. (a) acute-angled and scalene; 60°   (b) acute-angled and scalene; 80°

    (c) obtuse-angled and scalene; 35°   (d) acute-angled and isosceles; 55°

    (e) acute-angled and isosceles; 66°   (f) obtuse-angled and isosceles; 27°

    (g) right-angled and scalene; 130°

    (h) acute-angled and scalene; $x = 60°$; $y = 40°$

    (i) acute-angled and isosceles; $a = 75°$; $b = 30°$; $c = 105°$

(j) acute-angled and isosceles; $l = 65°$; $m = 65°$; $n = 115°$

(k) acute-angled and isosceles; $p = 70°$; $q = 70°$; $r = 40°$

(l) right-angled and scalene; $x = 55°$; $y = 35°$

3. (b) 8

4. $x = 50°$

5. 5 : 3

Exercise 38

1. (a) $a = 10$ (b) $a = \sqrt{74}$ (c) $c = \sqrt{21}$ (d) $c = 3\sqrt{3}$

(e) $b = \sqrt{39}$ (f) $b = 2\sqrt{5}$ (g) $c = 12$ (h) $c = 6$

(i) $a = 7\sqrt{2}$ (j) $b = 9\sqrt{3}$ (k) $a = 4\sqrt{2}$ (l) $b = 24$

(m) $a = 13$ (n) $c = 4$ (o) $b = 6$ (p) $a = 7\sqrt{2}$

2. (a) $6\sqrt{3}$ (b) $4\sqrt{3}$ (c) $5\sqrt{2}$ (d) $13\sqrt{2}$

3. (a) 500 km (b) $\sqrt{21}$ (c) 240 km

(d) $\sqrt{89}$ (e) $2\sqrt{11}$ (f) $10\sqrt{2}$

Exercise 39

1. (a) 2 400 cm² (b) 144 m² (c) 300 cm² (d) 1 200 cm²

(e) 64 m² (f) 14 m² (g) 75 cm² (h) $h = 20$ m ; 150 m²

(i) 16 m²

2. (a) 25 m² (b) 51 m² (c) 57 m²

3. (a) 12.5 (b) 6

4. $8\sqrt{2}$

5. $8\sqrt{3}$

6. 40 square inches

7. (a) 28 m (b) 32 m (c) 52 m

8. 10

Exercise 40

1. (a) 25π cm² (b) 49π km² (c) 36π m²

(d) 2.25π km² (e) 1.69π cm² (f) 6.25π m²

2. (a) 10π cm (b) 14π km (c) 42π m

(d) 3π km (e) $\frac{28\pi}{11}$ cm (f) 7π m

3. (a) $\frac{2\pi}{3}$ cm (b) 6π m (c) 100π mm

4. (a) 3π (b) 0.18π (c) 27π

5. (a) 120° (b) 40° (c) 270°

6. $\frac{(2\sqrt{2})}{\pi}$ 7. $\frac{(15 - \pi)}{2}$ 8. 324π

9. $s = 1$ 10. 2π 11. B $(-5, -3)$

12. N $(-4, 1)$ 13. 20

Exercise 41

1. (a) 36 cm^3 (b) 125 m^3 (c) $\frac{1}{2}\text{ m}^3$

   (d) $5\frac{1}{4}\text{ cm}$ (e) $\frac{1}{2}\text{ m}$ (f) $42\,000\,\ell$

2. 700 litres   3. 27 cans   4. 20 litres

5. $\frac{1}{4}$ 6. 60 cl

CHAPTER 4 Basic statistics

Exercise 42

1. (a) 18   (b) 36   (c) 6   (d) 1.4   (e) 11

2. (a) 15   (b) 7   (c) 15   (d) $\frac{(9x + 15y)}{(x + y)}$ (e) $3z$

   (f) $\frac{(5x + 8y)}{(x + y)}$

3. $\frac{(nx + y)}{(n + 1)}$ 4. 14.2   5. 40   6. $52 : 130$

7. 86   8. 9   9. 7

Exercise 43

1. (a) 7%   (b) 29%   (c) 67%

2. (a) 44   (b) 252   (c) only 1 learner (score of 60)

   (d) statement (i) only $(x < 10)$ (e) median $= x^1$

   (f) median $= x^1$ (g) $x = 16$ (h) 8

   (i) option (v)

Exercise 44

1. (a) 21 kg; 27 kg; approx. 68% of all observations lie within these limits

      18 kg; 30 kg; approx. 95.5% of all observations lie within these limits

      15 kg; 33 kg; approx. 99.7% of all observations lie within these limits

   (b) [115 km; 135 km]   [105 km; 145 km]   [95 km; 155 km]

   (c) 69.5 ℓ; 74.5 ℓ]   [67 ℓ; 77 ℓ]   [64.5 ℓ; 79.5 ℓ]

   (d) 1.025 ℓ; 1.075 ℓ]   [1.0 ℓ; 1.1 ℓ]   [0.975 ℓ; 1.125 ℓ]

2. 74   3. 10.5   4. 11 hours   5. Data pair (6, 6)

Exercise 45

1. (a) 0.167   (b) 0.125   (c) 0.28

2. 0.1875

3. (a) P(Spanish) = 0.05 (b) P(German) = 0.2 (c) P(German or French) = 0.35

Exercise 46

1. (a) (i) 15 (ii) 0.5
    (b) 28
    (c) 18%
    (d) (i) 0.167 (ii) 0.833 (iii) 0.75
    (e) 60 members
    (f) P(small/not hired out) = 0.8
    (g) 50%
    (h) (i) 0.5 (ii) 0.1 (iii) 0.6 (iv) 0.2
    (i) (i) 40 persons (ii) P(stayed/women) = 0.5
    (j) $\frac{11}{12}$N
    (k) (i) 0.667 (ii) 0.375 (iii) 0.625 (iv) 0.222

Exercise 47

Bar charts not shown.

(a) Honda = 22% ; Fiat = 12% ; Corsa = 38% ; VW Golf = 28%

(b) Pretoria = 10.7%; Kimberley = 18.7%; Johannesburg = 50.7% ; Durban = 20%

(c) Checkers = 38% ; Spar = 18.7% ; Pick 'n Pay = 43.3%

(d) Matric = 39.2%; Certificate = 10.8% ; Diploma = 18.8%; Degree = 31.2%

(e) 0 = 14% ; 1 = 34% ; 2 = 25.6% ; 3 = 20% ; 4+ = 6.4%

Exercise 48

1. (a) 840 (b) 1 (c) 30 (d) 35
2. (a) 720 different ways (b) 360 different displays
    (c) 56 different question combinations (d) 5
3. (a) 120 possible committees (b) 3 024 permutations
    (c) 420 different teams; probability = $\frac{1}{420}$

Exercise 49

1. (a) 0.005% (b) 1.495%
2. (a) $\frac{1}{4}$ or 25% (b) $\frac{7}{8}$ or 87.5%
3. 0.5

Solutions to pre-revision test

1. C $\frac{0.216}{0.54} = \frac{21.6}{54} = 0.4$

2. D $x = y + 10$

3. C $\frac{24}{1\frac{1}{4}} = 19\frac{1}{5}$ pieces; maximum = 19

4. D $\frac{5}{2} \times \frac{10}{1} = 25$

5. B (approx.) $60 \times \frac{1}{20} = 3$

6. C 1.5×40 mph = 60 mph;
       time $= \frac{80}{60} = \frac{4}{3}$ hours $= 1\frac{1}{3}$ hours

7. E LCM = 48: $\frac{34}{48}$ $\frac{24}{48}$ $\frac{18}{48}$ $\frac{36}{48}$ $\frac{27}{48}$
       Middle $= \frac{27}{48} = \frac{9}{16}$

8. B $\frac{90}{10} \times 110 = $ R990

9. C $8n + 3 + 1(n - 1) = 9n + 2$

10. C $(2\sqrt{7})^2 = 2^2 (\sqrt{7})^2 = 4 \times 7 = 28$

11. D $\left[\frac{K + (2K + 3) + (34K - 5) + (5K + 1)}{4}\right] = 63$
       $(11K - 1) = 252$ $K = 23$

12. B $x + y = 300$; $0.1x + 0.15y = 38$
       $x = 140$ g

13. B Total time = $4.2 \times 5 = 21$ min. Sum of 4 = 17. Difference = 4 min.

14. C From the cross-tabulation table: 30 enjoy both sports. Probability $= \frac{30}{60} = 50\%$

		Cricket		
Soccer	Enjoy	Don't enjoy	
Enjoy	**30**	15	45
Don't enjoy	5	10	15
	35	25	60

15. D Use Pythagoras' theorem – 3 : 4 : 5 triangle. Each side is doubled to 6 : 8 : 10.

16. A $x = -2$ and $x = \frac{3}{2}$, then $(x + 2) = 0$ and $\left(x - \frac{3}{2}\right) = 0$. Thus $(x + 2)(2x - 3) = y$
       $2x^2 + x - 6 = y$

Your test score (and recommended course of action):

≤ 8 correct answers	Poor	Requires serious revision of all areas of Basic Mathematics
9 – 13 correct answers	Fair	Requires selected revision of weak areas of Basic Mathematics
14 – 16 correct answers	Good	Requires general review to consolidate all areas of Basic Mathematics

Solutions to post-revision test

1. A

$\frac{1}{2} + \frac{1}{4} + \frac{1}{8} + \frac{1}{8} = \frac{15}{16}$ (removed) (or $\frac{1}{2^4}$ air *left* after 4 strokes)

2. C

$\frac{3\frac{1}{2}}{\frac{1}{2}} = 7$ cups/can; need min. $\frac{60}{7} = 8\frac{4}{7}$ cans = 9 (rounded)

3. B

$\frac{(4-5)}{100} = -\frac{1}{100}$

4. D

$18 + 0.1d = 25 + 0.05d$ $d = 140$ km

5. B

A = 3 bolts/s; B = 5 bolts/s; A + B = 8 bolts/s; $\frac{200}{8} = 25$ s

6. B

Apples (6): Peaches (1): Grapes (2); Sum = 13; difference $\frac{(6-2)}{13} \times 78 = 24$ kg

7. A

Circle = $2\left(\frac{22}{7}\right) = 6\frac{2}{7}$; equilateral triangle = $4\sqrt{3}$; scalene triangle = 6

8. B

LCM = 60; $\frac{45}{60}$ reduces to $\frac{11.25}{15}$

9. D

$0.05x + 0.20y = 0.15(x + y)$ with $x = 3y$ (grass seed B) = 6 kg

10. C

Car speed = 1.3 × 50 = 65 km/h. Distance (car) = 65 × 4 = 260 km.
Distance (truck) = 50 × 4 = 200 km. Difference in distance = 260 – 200 = 60 km.

11. E

Cereal	:	Bacon	:	Eggs	
$\left(\frac{1}{3}\right)\left(\frac{5}{4}\right)$	:	$\frac{5}{4}$	:	1	
$\frac{5}{12}$	:	$\frac{5}{4}$	:	1	$\frac{12}{5} = 240\%$

12. B

$x = 1\,000 \times 0.01 = 10$; $y = \frac{1}{0.01} = 100$; $z = 100(1 - 0.9^2) = 100(0.19) = 19$

13. E

The median lies between the third and fourth numbers, which are all multiples of 8 (i.e. 48 and 56).

The fifth and sixth numbers are therefore 64 and 72, respectively.

Highest = 72.

14. D

$\frac{(45 + 35 - 30)}{60} = \frac{50}{60} = 83\%$

Application of the addition rule of probability.

Soccer	Cricket		
	Enjoy	Don't enjoy	
Enjoy	**30**	15	**45**
Don't enjoy	5	10	15
	35	25	60

15. D

$2l + 2w = 42$ $\qquad l = 2w$ $\qquad w = 7$ cm $\qquad l = 14$ cm

Diagonal: use Pythagoras' theorem

16. B

$(x + 2)(x - k) = y$ $\qquad x^2 - kx + 2x - 2k = y$

Substitute (0, 4) $\qquad$ Solve for $k = -2$.

Equation is $y = (x + 2)(x + 2) = (x + 2)^2 = x^2 + 4x + 4$

Your test score (and recommended course of action):

< **8** correct answers	Poor	Much more serious revision required of all areas of Basic Mathematics.
9 – 13 correct answers	Fair	Areas of weakness remain. Requires further selected revision of weak areas.
14 – 16 correct answers	Good	You are ready for the GMAT test. Now focus on GMAT test questions.

---ooOoo---

Good luck with your GMAT: